Monte Verità Proceedings of the Centro Stefano Franscini
Ascona

Edited by K. Osterwalder, ETH Zürich

Field-Scale Water and Solute Flux in Soils

Edited by
K. Roth
H. Flühler
W.A. Jury
J.C. Parker

1990

Birkhäuser Verlag
Basel · Boston · Berlin

Editors' addresses:

Dr. K. Roth
Department of Soil
and Environmental Sciences
University of California
2208 Geology
Riverside, California 92521
USA

Prof. H. Flühler
Bodenphysik ETHZ
ETH-Zentrum
NO H46
8092 Zürich
Switzerland

Prof. W.A. Jury
Department of Soil
and Environmental Sciences
University of California
Riverside, California 92521
USA

Prof. Dr. Jack C. Parker
Center for Environmental
and Hazardous Material Studies
Virginia Polytechnic Institute
241 Smyth Hall
Blacksburg, VA 24061
USA

Deutsche Bibliothek Cataloging-in-Publication Data

Field scale water and solute flux in soils / ed. by K. Roth ... – Basel; Boston; Berlin: Birkhäuser, 1990
(Monte Verità)

ISBN-13: 978-3-0348-9969-7 e-ISBN-13: 978-3-0348-9264-3
DOI: 10.1007/978-3-0348-9264-3

NE: Roth, Kurt [Hrsg.]

Softcover reprint of the hardcover 1st edition 1990

Printed on acid-free paper

CONTENTS

Field-Scale Water and
Solute Flux in Soils
Monte Verità

INTRODUCTION

The soil profile forms the interface between the atmosphere and the geosphere. Material and energy fluxes through the soil profile therefore are of interest to a variety of scientific disciplines, as well as to individuals concerned with land and water resource management. Because water and nutrients vital to plant growth and crop yield flow through the surface soil profile, the study of material and energy fluxes through soil is important to the agriculture and food production industries. However, because the soil profile overlies the ground water resource, flow of chemical contaminants through the soil is of importance to a variety of scientists, agencies, industries, and private citizens concerned with water pollution and resource management.

Quantitative characterization of material and energy flows through soil is extremely difficult, because the properties of the soil profile that are important in transport are quite variable in space, and can change significantly over time. Moreover, spatial heterogeneity is manifest at all distance scales, so that measurements of transport properties are difficult to interpret or to average. This heterogeneity also makes transport modelling difficult, both because not all of the processes that contribute to the transport of matter and energy through soil are well understood, and also because measurement limitations greatly restrict the spatial scale at which the transport processes can be formulated.

This book contains the proceedings of the first workshop held at the Monte Verità resort near Ascona, Switzerland on September 24–29, 1989. It is intended to represent a cross section of current research on water and solute transport through soil, as well as group reports on four current areas of interest in transport. The first part of the book consists of the reports prepared by the four Think Tank discussion groups. These groups, which were formed from the body of the conference attendees, held daily meetings during the conference to discuss issues surrounding the topic assigned to the group by the conference organisers. Each group defined its own format, and developed a unique response to the charge of developing a summary report defining its activities. The second part of the book contains a selection of contributions that were presented at the workshop.

Some special circumstances as well as the hard work of many people were responsible for the success of the workshop and the production of this book. The most important circumstances behind the success of the workshop were the marvellous setting of the Centro Stefano Franscini and the generous financial support provided by the Swiss Federal Institute of Technology (ETH). The running of the workshop was facilitated greatly through the efforts of the Soil Physics group of the ETH. We owe a special debt of gratitude to Dr. Bernhard Buchter, Markus

Flury, Thomas Gimmi, Sabine Koch, Hanspeter Läser, Andreas Papritz, Gérald Richner, and Martin Schneebeli for their efforts preparing for the gathering, and working behind the scenes during the week, and to Flavia Crameri for patiently typing many manuscripts into the proper format. We also acknowledge an important, albeit anonymous contribution from the many outside reviewers who provided valuable comments on the research papers in this book. Finally, we express special gratitude to the Director of the Centro Stefano Franscini, Dr. Konrad Osterwalder, for his active support during the planning, execution, and preparation of proceedings for this workshop.

Zürich, Switzerland
Riverside, California USA
Blacksburg, Virginia USA
September 1990

Kurt Roth
Hannes Flühler
William Jury
Jack Parker

Part 1

Think Tank Reports

Field-Scale Water and
Solute Flux in Soils
Monte Verità

FLOW AND TRANSPORT MODELING APPROACHES: PHILOSOPHY, COMPLEXITY AND RELATIONSHIP TO MEASUREMENTS

Discussion leaders : A. Rasmuson and H. Flühler

Participants : T. Addiscott, G. Dagan, T. Gimmi, P. Hufschmied, D. Imoden, S. Koch, G. Kachanowski, G. Karlaganis, P. Lachassagne, D.O. Lomen, P.S.C. Rao, B. Sagar, R.E. White, P. Wierenga, J.H.M. Wösten

Defining the Goal

Models for flow and transport in the geo- and pedosphere are developing at a rapid pace. This field of research is receiving considerable attention and increasing support based on several **implicit assumptions**. It is for instance assumed that such models are or will become **regulatory** as well as **research tools** for interpreting cause and effect relationships in environmental pollution events. Furthermore, it is assumed that such models can ultimately be **scaled up** to represent the significant features of reality, that is, a field soil or even a catchment with its space and time domain, including all relevant features and elements such as structure, biota, liquid, and mineral phases, etc. The choice of the workshop topic and the outcome of this meeting indicate that these implicit assumptions are not entirely based on solid grounds.

The **goal** of this think tank was to distill **ideas** and **personal views** from the on-going discussion of what models are and what they are presumed to be; ideas and personal views which are **relevant** for enhancing our **understanding** of field scale transport phenomena and **useful** in a regulatory and management sense.

The essence of our group discussions can be summarized by a collection of **questions** which floated to the surface of our "tank" during the initial round:

Do we need more models? How can we make models relate better to the real world? How precisely do we need to know parameter values - and do we have the sensitivity analysis that will enable us to decide? In estimating model parameters should we scale up from smaller to larger systems or should we go straight to the larger system? How do we establish safety factors for soil contaminants for regulatory purposes? What can we do, other than measuring concentrations in the soil, to assess the safety of the soil? How many samples are needed to quantify adequately the amount of a contaminant in, or its flux from an area of land? What models, existing or new, do we need to help with environmental protection? And what level of complexity and what scale are needed? How can we avoid the mistakes made elsewhere with the use of models for regulatory purposes?

Model Philosophy

It was apparent from the discussions in the think tank that the philosophy of modelling is a difficult subject with many aspects, but the following structure gradually evolved.
The number of different models is probably comparable with the number of modelers. Nevertheless, we tried to distinguish different **categories** to classify field scale models:

- purely deterministic vs. stochastic models
- stochastic vs. deterministic models with spatially variable parameters
- black-box models which produce a system response to an input signal vs. models with a variable degree of process resolution
- functional vs. mechanistic models
- physically based vs. empirical models
- research vs. management models
- rate vs. capacity models

These model categories are judged on different levels and therefore represent at best fragments of a larger classification scheme. Various hybrid categories may be defined by combining the above or additional criteria. The limited discussion time allowed us to categorize field scale models, but prevented us from getting bogged down in terminology.
The **(external) requirements** which a model must meet and the **(internal) model requirements** which must be satisfied for using a particular model are not independent. It is obvious that the former should be of prime concern but the priorities given to the two sets of requirements are occasionally mixed up. Any model, in particular a field scale transport model, should be designed to satisfy explicitly defined (external) requirements, i.e. with regard to its **objectives**, its **spatial** and **temporal scale** and its **degree of resolution**.

Model Objectives

A model can rarely be categorized in isolation from its objective. The broad category "research or management" is obviously not a rigid distinction. The latter may develop from the former or more seldom vice-versa. The modelers' motivation and preferences are often not taken into account or even tacitly denied when a model is being advertised to meet the needs of society. More realistically the model objective is often tailored to the needs of the funding agencies or to the modelers own experience. The majority of the discussion group agreed—some disagreed—that soil physicists and other scientists active in this field have too often not asked why—other than for the scientific thrill of doing it—they have been doing what they have. Hence, the definition of model objectives encompasses widely diverging views.

Model Scale and Degree of Resolution

For the modeler the **scale** and the **degree of resolution** is what the "ground area on the purchased lot" and the "structure" is for the building constructor. **Model scale** has no degrees of freedom and does not allow adjustments or choices for a particular objective or application. It is a defined feature of the given system. Model scale has many implications i.e. the measurability of parameters and variables. The **degree of resolution** on the other hand should – and could in some cases – be tuned to include (only) those processes and system noises which are essential to reproduce the pertinent features of reality at the scale considered. It may be broken down in several ways like

- temporal and spatial resolution
- definition of subcompartments
- process resolution at various levels (lumping or splitting), which eliminates or adds certain variables and parameters
- accuracy of parameter values and variables
- accuracy and completeness of the test data base, etc.

The scale of the discrete elements in the model space should correspond with the **scale** of at least some of the **measurable properties**. The finer the splitting into multi-processes or multi-sites and the higher the grid **resolution** in space and time, the more important it is to quantify the **system's heterogeneity**. In both the model and the real world, the variability needs more and more attention as the degree of resolution increases. In this context, it is valuable to distinguish between the variability of **extrinsic** (externally imposed) and **intrinsic** (internal) properties and processes . The spatial distributions of input functions (extrinsic property) may dominate the spatial variance structure of a given soil property. In some cases they are known or can be more easily measured than the resulting distributions of concentrations deeper down in the soil (intrinsic property). Examples are spray bar strips, wheel tracks, or in natural habitats stemflow or canopy throughfall patterns. Examples of intrinsic variability are the layering of parent rock material, i.e. in alluvial or colluvial soils. **Noisy initial** or **boundary conditions** may be less significant in systems, which can be described with parabolic differential equations. Spatially and temporally **variable transmission properties** which may depend upon spatially and temporally varying state variables are one of the data base deficiencies making high resolution models untestable. In addition, the **variability of processes**, both spatial and temporal, is possibly an even more crucial gap in our knowledge than the lack of information about the distribution functions of variables and parameters.

Different applications require different primary answers and therefore different **accuracies** of model predictions, of model inputs and of the various elements in the data base. For example, for a substance of high toxicity, like a pesticide, the time of first arrival may be the key output. For non-toxic substance like fertilizer, the flow of mass to the groundwater may be of main interest. These differences will affect the accuracies needed in model prediction.

Model Validation

The problem of model validation was discussed at length. Even though a large number of transport models is available, few have been adequately tested. Model validation in its rigorous and narrow sense requires a model to be run with completely independently determined system parameters, a prerequisite which is rarely met in field case studies. It was emphasized that any level of model validation, stringent or just tentative, depends on

- the **purpose** and therefore type of model
- the **model scale** (soil cores, lysimeters, fields, catchments, regions, and time scale)
- and especially upon the **target solute** (toxic substance, fertilizer, heavy metals)

The rigor of model validation needed should depend on **cost-benefit analyses** or **risk assessments**. These approaches should pose the questions like how much additional effort should be invested in data collection or modeling in relation to the perceived increase in accuracy of model predictions. It should be kept in mind that the costs of modeling are often small compared with the investments needed to obtain a sufficient field data base.

Recommendations and Conclusions

1. Available Data Base

It was apparent from our discussions that data from a fairly large number of field studies could be compiled. However, in most cases the comparison data sets necessary as model inputs are largely lacking. This is especially true for the validation of stochastic models which require knowledge of the spatial variability structure of parameter values. In only a few studies is the spatial variability of transmission and storage properties and local flow patterns being measured. Such measurements may or may not reflect the effective state of the system and often do not have the appropriate length and time scale. It was also noted that average values and spatial structures for various chemical and biological processes are lacking in most cases.

It is therefore proposed that a coordinated set of field experiments be conducted at different locations to characterize **unsaturated, transient water flow and solute transport** as for instance the "Las Cruces trench study", the "Riverside field scale tracer trials", or other experiments reported recently in the literature (Wierenga 1990, Jury 1990). These studies should be designed to provide data suitable for testing deterministic and stochastic models. Detailed

protocols need to be developed for design and conduct of such studies equivalent to the studies carried out in the saturated zone (Borden site, Cape Cod, Columbus Air Force Base 1990).
Joint ventures of field experimentation would be an incentive for modelers to consider more carefully the implications of small scale, imprecise and expensive data sets needed for their modeling attempts.

2. Guidance for future Studies

A **protocol** for **future studies** should be developed so that the data collection is suitable for model testing. It is therefore proposed that

- a **minimum** and an **optimum data set** required for model testing be defined
- model **validation** and **performance criteria** be developed
- **synthetic data sets** be generated by comprehensive models to test simpler models.

Some of our thinking in this direction is summarized below.

3. Synthetic Data Sets

The discussion about the potential of synthetic data sets was controversial. It was emphasized that testing a simpler model with a more comprehensive model is permissible if and only if the simpler model is a subset of the latter. This approach presupposes that the comprehensive model is valid. In one instance (Nicholls et.al, 1982) a "simpler" model simulated the movement of chloride and pesticides at least as well, if not better, than a "comprehensive" model.

4. Criteria for Model Validation

The **criteria for model validation** should be specific for the **solute** of interest and the **scale** of the model. It should be clearly stated whether a particular model is intended to be used for estimating

- the **earliest** possible time of arrival (e.g. for highly toxic solutes),
- the **flow of mass** (e.g. NO_3-leaching).
- the **mass** of solutes **remaining** in the soil (e.g. plant nutrients)
- the **maximum** (average) **outflow concentrations** (e.g. drinking water standard for EEC)

The estimation of mass flow might require a less stringent **accuracy** of prediction than the prediction of the earliest possible arrival time of a highly toxic compound. Required accuracy may be related to a cost-benefit analysis of the target solute.
If the concentration of a hazardous solute within the **reception compartment** such as the aquifer or surface waters is required to be below a given standard, then the mass of the solute being displaced through the soil should be related to the **receptor volume.**

5. Data Set Requirements

The **minimum or optimum data set requirements** are focused on model validation. Field experiments used for model testing must provide a **minimal data** set which may be **categorized** in different ways:

- **A priori knowledge** which may easily be obtained such as climatological and pedological information, and **measurements** which can be carried out with soil core samples.
- Another structure for the minimal data set is the distinction between climate, soil, plant, management and target solute.
- A third categorical structure for data relates to location and time of observations, that is the water and solute mass fluxes at the upper and lower boundaries and the spatial distributions of time-dependent variables at selected times (initial boundary conditions and subsequent depth distributions of the variables).

6. Minimal Data Set

The sampling strategies of large scale field experiments are often tailored to meet fairly well defined questions or to input requirements of certain types of models. Therefore, in many if not most cases, much very basic information is either not obtained or not reported. We propose here a minimal data set to stimulate discussions along these lines and to encourage scientists to report more complete data sets which allow a greater flexibility in model validation. However, we realize that the number of variables measured depends on the objectives of the study, the models chosen as well as on other constraints such as funding and experimental experience.

A minimal data set should contain the following information :

Climate

- precipitation measurements (amount and, if possible intensities)
- minimum and maximum air temperature
- inputs for evaporation calculations

Soil

- soil description (as used for defining soil series) from a soil characterization information system (site, location, latitude and longitude, morphological features)
- geological and hydrogeological information
- topography and elevation
- transmission and storage properties for water
- soil structure (density, pore and particle size distribution, indirect structural parameters such as infiltrability)

Plant
- crop type, yield-nutrient relationships
- depth of rooting zone

Management
- cultivation methods, fertilizer and manure applications
- drainage, irrigation and harvesting methods

Target solute
- method of application
- chemistry and biology of the solute, relevant transformation and soil retention characteristics

Solute specific variables of the minimal data set are

<u>nitrate</u>
- soil organic N
- rate constant of mineralization

<u>pesticides</u>
- half life time in soil
- optimum water content partition coefficient (solubilities)
- organic carbon content
- toxic degradation products
- volatility

<u>heavy metals</u>
- cation exchange capacity
- pH
- organic ligands (total organic carbon (TOC) or more specific analyses) and other organic phases
- solution composition

<u>phosphate</u>
- pH
- clay content (or soil P-sorption capacity obtained from batch experiments)

<u>hydrocarbons</u>
- organic carbon content
- solubility
- volatility

<u>bacteria, particulate matter</u>
- type of organisms (size,shape,species)
- type of particles (size,shape,charge-if any)

Some of the properties mentioned in this provisional list are not well defined and cause substantial methodological problems. A careful documentation of how such properties were measured is essential for any kind of data set especially in those cases when data are handed over to researchers not involved in that particular experiment or lacking experience in field experimentation..

7. Outlook

More **interaction** between **modelers** and **experimentalists** is recommended. Such interaction would benefit from international studies of the type INTRAVAL, a Swedish nuclear repository project (1987). In this study a set of laboratory and field experiments is interpreted and modeled by different project teams using different modeling approaches. It was suggested that the Monte Verita Workshop may serve as a platform along these lines.

The value of **high quality field information** should receive a **more favorable rating** in the minds of the reviewers of journals and funding agencies. Ideally, high quality field experimentation and competent modeling should go together but, in reality, most projects are stronger in one or the other respect. The apparent overrating of model oriented contributions which require in many cases less investment of time and funds than obtaining the needed field data base, should be re-examined by agencies.

References

INTRAVAL Project Proposal: *Swedisk Nuclear Power Inspectorate (SKI)*. Report **87**:3, 1987.

Freyberg, D.L. 1990: *The Borden Field Experiment: Transport and Dispersion ofTracers and Organics*. In: Murarka, I.P. and S. Cordle (Eds.) Electric Power Research Institute EPRI EN 6749 pp.12/1-12/23.

Nicholls, P.H., Bromilow, R.H., and Addiscott, T.M., 1982: *Measured and simulated behaviour offluormeturon, aldoxyearb and chloride ion in a structured soil*. Pesticide Sci. **13**, 475-483.

Wierenga, P.J., D.B. Hudson, R.G. Hills, I. Porro, M.R. Kirkland, and J.Vinson 1990: *Flow and Transport at the La Cruces Trench Site: Experiments 1 and 2*. NVREG report CR-5601, pp. 413.

Young, S.C. and J.M. Boggs, 1990: *Observed Migration of a Tracer Plume at the MADE Site*. In: Murarka, I.P. and S. Cordle (Eds.) Electric Power Research Institute EPRI EN 6749 pp. 11/1-11/18.

Field-Scale Water and
Solute Flux in Soils
Monte Verità

EFFECTIVE LARGE SCALE UNSATURATED FLOW AND TRANSPORT PROPERTIES

Discussion Leaders: J.C. Parker and L.W. Gelhar

Participants: R. Ababou, G. Butters, G. de Marsily, D. Mulla, M. Schneebeli, F. Stauffer

Decisions concerning the management of subsurface water resources requires information on the behavior of water and chemicals in soils and groundwater at relatively large scales, i.e., fields, small watersheds and large hydrologic basins. Numerical models are often used to provide such information. However, practical limitations on data collection always impose constraints on the resolution which can be obtained in the distribution of model parameters in space. Under the best of circumstances, direct measurements of soil properties are available at only a small number of locations relative to the number of nodes in the numerical mesh and correspond to measurement volumes generally much smaller than nodal (or element) volumes. Thus, field scale numerical simulations almost always invoke some scale-up process—implicitly if not explicitly. Many of the papers in this conference relate to this problem. Most papers deal in some fashion with assessment of the variability of soil hydraulic and transport parameters, extensive variables and/or mass fluxes or of methods of dealing with these problems in models. Several papers directly address the use of "effective" parameters in field scale models based on simple empirical averaging rules or on more rigorous theoretical protocols. With regard to the description of unsaturated flow, it is generally contended that the Richards equation provides a satisfactory description of water flow under conditions of negligible gas phase impedance at a scale which is sufficiently large to encompass pore scale variability but not so large as to suffer from larger scale heterogeneity effects. In the case of porous media with continuous fractures or other channels, it may be noted that unanimity of opinion does not exist whether such a range may be defined to enable application of the Richards equation to the porous medium as a single continuum.

Unsaturated soil hydraulic properties (water conductivity and water capacity) are highly nonlinear functions which are known to exhibit marked spatial variability in geologic media at various scales of observation. For a specific simulation problem, certain components of this variability may be practical to treat deterministically (e.g., specific stratigraphic units or pedologic layers), while others may be addressed explicitly only within a stochastic framework since measurement frequency always imposes limitations on the level of detail known exactly. One approach to the problem of nondeterministic soil heterogeneity is to carry out Monte Carlo simulations of multiple realizations of the stochastic system with properties defined at a fine scale of resolution. Such an approach is, however, generally unrealistic due to the excessive computational cost of repeatedly solving a large system of highly nonlinear equations associated with heterogeneous systems on a fine grid.

Regardless of the methodology employed to deal with heterogeneity, for reasons of computational efficiency it is desired to utilize as coarse a numerical grid as possible without introducing excessive numerical dispersion in the results. "Effective" soil properties provide a mechanism for dealing with the two-edged problem of heterogeneity as well as discretization-dependent numerical dispersion. Effective flow properties may be operationally defined as those which when employed in the conventional flow equations at a desired (i.e., relatively coarse) level of discretization emulate the bulk response of the heterogeneous soil in some average sense.

In petroleum reservoir modeling, a stepwise scale-up approach has been described by Kossack et al. (1989). The approach involves a hierarchical description of the geologic system from small to large scale in which geologic features which are detectable (e.g., via drill logs or geophysical investigations) are distinguished explicitly while finer scale variability is treated stochastically. Beginning with the fine scale features, fine-grid, small-scale subdomain simulations are conducted for multiple realizations of the stochastic properties and average conductivity and capacity functions are calculated by averaging over the subdomain to define "pseudo-properties" at this scale. These pseudo-properties are then employed in larger scale simulations to define pseudos at the next larger scale. Such an approach is flexible and enables automatic accomodation of deterministric as well as stochastic variability in addition to providing corrections for grid-dependen numerical dispersion. Limitations of the methodology arise due to the inherent grid-scale dependence of the properties which requires recalibration for significant changes in discretization. More seriously, the method has limited generality when applied to problems involving boundary and initial conditions dissimilar to those used in the pseudo-curve calculations. This occurs because time-dependence of the pseudo-properties is generally disregarded in this approach. A more theoretically sound approach to deal with the stochastic analysis problem, which enables incorporation of time-dependent effect, involves employing a perturbation approach to describe the mean flow equation. Under certain simplifying assumptions regarding the local form of the conductivity and capacity functions and their statistical properties, analytical expressions for effective properties may be developed (Yeh and Gelher, 1983; Yeh et al., 1985; Mantoglou and Gelhar, 1987). Alternatively, less restrictive assumptions may be invoked and the form of the effective properties may be determined numerically. The result obtained by both the analytical and numerical approaches indicate that large scale effective properties exhibit hysteretic and anisotropic behavior which depends in a rather complex manner on time and space derivatives of the mean head. Laboratory experiments (Stephens and Herrmann, 1988), field observations (McCord and Stephens, 1987) and numerical simulations (Polmann et al., 1988) demonstrate some effects of large anisotropy in heterogeneous soils. An important feature of such methods is the feasibility of computing the variance of predicted head, water contents or fluxes at any location and time in addition to the expected value. The stochastic methodology is at present limited to domains considered as piecewise statistically homogeneous. Subdomains having deterministic boundaries may be considered based on stratigraphic or other qualitative data, but averaging across deterministic features is not feasible in this context nor is assessment of changes in properties needed to permit less stringent grid spacing. To date, no work has addressed the effects of small scale hysteresis and anisotropy in hydraulic properties on large scale behavior nor have interactions between stochasticity in boundary conditions (in time and space) and spatial soil variability been considered.

Calibration of unsaturated flow models to account for soil heterogeneity should begin with as detailed a qualitative description of the stratigraphy and geometry of the site as possible so that known determinisitic features may be treated. Within distinguishable soil zones, information will be needed on means, variances and correlation length scales of local scale hydraulic conductivity-water content-capillary pressure relations. This will generally entail the use of soil cores to determine water content-capillary pressure relations in the laboratory. It may be noted that correspondence between laboratory and field measurements of water retention characteristics is sometimes poor. However, field measurements also exhibit a high measurement error and are much more tedious to perform. Determination of unsaturated hydraulic conductivity is much more difficult and it has become common practice, at least in the US, to make direct measurements of only the saturated conductivity and relative permeability estimate from the moisture retention curve. Such procedures are rather approximate and need to be validated before they can be used in field applications. A substantial need exists for rapid field methods of estimating unsaturated soil hydraulic properties. Realistic data requirements for hydraulic property measurements would be 10-30 soil core determinations to define means and variances of required properties. Since many more samples would be required to estimate correlations scales accurately and since flow predictions are relatively insensitive to this parameter, estimates of this property based on experience with larger data sets is reasonable in practice. For stochastic models of transport, no additional information is needed to describe the dispersive process although model sensitivity to correlation scale in the direction of the mean flow becomes much greater and more attention will need to be given to the development of protocols for obtaining such information.

References

Kossack, C. A., J. O. Aasen, and S. T. Opdal. 1989. Scaling-up laboratory relative permeabilities and rock heterogeneities with pseudo functions for field simulations. Proceedings, Tenth SPE Symposium on Reservoir Simulation, February 6-8, Houston, p. 367-390.

Mantoglou, A., and L. W. Gelhar. 1987. Stochastic modeling of large-scale transient unsaturated flow systems. Water Resour. Res., 23(1), p. 37-46.

McCord, J. T., and D. B. Stephens. 1987. Laeral moisture movement on sandy hillslope in the apparent absence of an impending layer. Journal of Hydrological Processes, 1(3): 225-228.

Polmann, D. J., E. G. Vomvoris, D. B. McLaughlin, E. M. Hammick and L. W. Gelhar. Application of stochastic methods to the simulation of large-scale unsaturated flow and transport. U.S. Nuclear Regulatory Commission, Report#NUREG/CR=5094, Sep 88.

Stephens, D. B. and S. Herrmann. 1988. Dependence of anisotropy on saturation in a stratified sand. Water Resources Research, 24(5):770-778.

Yeh, T.-C. and L. W. Gelhar. 1982. Unsaturated flow in heterogenous soils. Proc. AGU Symp., Role of the Unsaturated Zone in Radioactive and Hazardous Waste Disposal, Ann Arbor Science, 71-79.

Yeh, T.-C., L. W. Gelhar and A. L. Gutjahr. 1985. Stochastic anlysis of unsaturated flow in heterogeneous soils. 2. Statistically anisotropic media with variable alpha. Water Resources Res., 21(4):457-464.

Field-Scale Water and
Solute Flux in Soils
Monte Verità

EVALUATION OF FIELD PROPERTIES FROM POINT MEASUREMENTS

Discussion Leaders: H. Selim and R. Schulin

Participants: J. Böttcher, M. Braun, B. Buchter, C. Gascuel, M. Kutilek, C. Lin, D. Myers, A. Rinaldo, D. Russo, R. Webster

Goal: To identify soil properties which influence water and chemical movement and interactions in soils on the field scale and to identify parameters (e.g. solute concentrations) for ecosystem analysis (e.g. balances) on the "problem" scale (e.g. field, catchment, region).

A general problem in determining fluxes on a field scale is to estimate values for large areas (say 1 ha to 100 ha) from measurements or outputs from models based on measurements made on small areas (typically 50 cm^2 to 1 m^2). Kriging is now recognized as the most generally reliable method of local estimation, either for points (areas of the same size and shape as those on which the measurements were made) or blocks (areas of side or diameter no larger than that over which the variogram is reliable). For larger areas locally kriged estimates can be averaged to give values with smaller estimation variances than would be obtained by classical methods.

We should also recognize that at the field scale variation can be anisotropic and take such a possibility into account when planning a sampling scheme. We should also take into account the temporal dependence of parameters in physical models and parameters of biological processes.

1. Type of Problem

Field scale models usually employ "effective field properties" which are related to local physical properties only implicitly through the specific structure of the model. Outside the context of the model these parameters, therefore, have in general no or little physical significance (e.g. effective hydraulic conductivity). True prediction of "effective field properties" from point measurements requires well-defined explicit procedures such that the physical meaning of these proper-

ograms. Use of other variables to calculate cross-covariance functions, cross-spectra, phase-spectra, and coherency-spectra may be very helpful to identify interactions between different parameters. Geostatistical procedures can be applied in principle for interpolation purposes provided that the variogram and the drift can be properly estimated.

2. Types of Variables

With respect to modeling, three basic kinds of soil variables may be distinguished: input, output, and related variables. Input variables refer to soil properties and boundary conditions needed to specify a model for a given case. Soil properties may be parametrized either as basic parameters such as hydraulic conductivity or as functional relationships e.g. between soil properties and state variables such as $K(\theta)$ or $\psi(\theta)$ relationships. Output variables refer to state veriables which are calculated as model outputs such as flux or concentration of a chemical. Comparison of measured and predicted output variables is the basis of model calibrationRelated variables refer to auxiliary information such as texture or soil morphology which is not directly used as model input. It is recommended that measurements onfield soils are identified in terms of these types of variables.

Prior to a measurement campaign, two questions pertaining to the measurement of soil variables or characteristcs have to addressed:

a) Whichvariables shall be measured and by which methods?

b) Shall the data be archived and made available to other researchers?

Soil properties are either measured on samples removed from the soil profile or in situ in the in the field. Taking measurements from soil samples in the laboratory is appropriate if the variables under study are not affected by the sampling procedure and do not significantly change under laboratory conditions until completion of the measurement. This is the case for some soil physical properties such as texture as well as many soil chemical properties such as pH, cation exchange capacity, exchangeable acidity, organic carbon content which are important factors of the chemical retention capacity of a soil. Although they can be measured most easily and accurately in the laboratory, some of them such as exchangeable acidity or organic carbon content are often not determined, however. Soil moisture content and bulk density are often determined gravimetrically, which by necessity involves destructive sampling. Although in general less sensitive than gravimetric methods, so called "non-destructive" methods such as the neutron attenuation technique or time domain reflectometry are better suited for monitoring changes in soil moisture content, as they allow repetitive measurements at the same location

Measurements of flow and transport parameters such as hydraulic conductivity, however, are strongly dependent on soil structure. The latter is not adequately represented in small samples and in addition also disturbed by the sampling procedure. Therefore, input parameters for flow and tranport models must in general be evaluated *in situ.* whereas the response or output variables such as concentration profiles over depth may be evaluated from cores at the laboratory.

Because of their sensitivity to environmental factors and their ability for rapid change, kinetic parameters of microbial reactions (e.g. denitrification) have to be evaluated from field measurements.
If the data are to be archived and made available to other researchers, it is important to document the methods of measurement and recording as well as to determine and to document any relevant auxiliaryinformation.

3. Calibration and Validation

The choice of soil variables to be measured is not only affected by the aims of the analysis but also by difficulties in collecting and recording the data. Due to physical or economical barriers,it may be necessary to use proxy variables. Such variables will require calibration in some form. Although having a physical basis for this calibration, would be in general desirable, using an empirical relationship may be all that is possible. In many cases the scale of measurement is the crucial aspect of this relationship.
Direct methods for the determination of transport characteristics such as saturated and unsaturated hydraulic conductivity are usually time consuming, labourious, and thus, expensive. Consequently, such methods are rarely used if many data have to be collected to assess temporal and spatial variability. Less direct, but much more rapid methods based on simplifying assumptions such as infiltration tests then apply although they lead to cruder and less precise results. Instead of predicting hydraulic parameters from correlations with other soil parameters such as soil texture, it is recommended to combine the results of different flux experiments using optimization procedures. For example, the functional form of the conductivity curve may be determined from laboratory measurements on core samples and used in combination with the cumulative infiltration curves to optimize field saturated conductivity values. Monte Carlo simulations can be used to assess the sensitivity of such an approach.
Generally, estimates of soil hydraulic characteristics should be first tested against simple field flux tests before they are used for statistical analysis of soil spatial and temporal variability or as parameters in more elaborate models.

4. Optimal Estimation and Interpolation: Some Pitfalls of Kriging

There is no one form of kriging, and the particular form of kriging chosen should be determined by what is already known about the variable and its behaviour in space. Each form of kriging has its own assumptions, which if satisfied lead to minimum variance estimates. All assume spatial dependence at the scale of sampling. If there is none then classical methods of estimation apply, and interpolation can do no better than return the mean of the data everywhere.
The quality of kriging estimates depends on the quality of the variogram on which it is based. A well estimated variogram requires a fairly large sample. As a very rough estimate, the minimum sample size should be 100 data for one dimension and 400 data for two dimensions if anisotropy

has to be taken into account. Determination of a continuous variogram function from these data requires the choice of a suitable variogram model which can be fitted to the experimental data. The choice of the variogram model as well as of the fitting procedure are by no means straightforward still a matter of controversy and research. One possibility is to use weighted least squares approximation. The fit of different models may be compared either using the residual sum of squares or the Akaike Informatio Criterion. Validation studies have shown that this works well.

Since the variogram isneither theoretically nor practically unique, so neither does kriging provide a unique estimate for any point or block. Experience shows, however, that kriging is rather robust with respect to the choice of the variogram model. Kriging estimates obtained with different reasonable variograms are closely similar. And although kriging variances depend more sensitive on the variogram than the kriging estimates, they are generally less than the errors incurred in other methods of interpolation.

In many cases, the question of how the user has implemented an interpolation scheme is more important than the choice of the scheme (e.g. kriging, inverse-distance weighting, splines). In the case of kriging, the geometric configuration of the estimated points or blocks and their supports may completely override the effect of other choices such as the variogram model. One should not have to take an author's word that the results are valid; enough information should be supplied to allow others to reproduce the results.

Results are often given graphically, for example by contoured plots. The packages that produce such images may have their own characteristics and are highly dependent on the nature of the input. For example, they depend on the mesh of the grid of points used for contouring. One should not treat such plots as truth. They almost never are, and they may suggest or imply conclusions not clearly supportable by the data.

5. Sensitivity to Measurement Errors and Artefacts

Model parameters should determined by direct measurements whenever possible. Unfortunately, as has been pointed out before, this is not always feasible. Key parameters of soil transport models such as the hydraulic conductivity must in general be determined as functions of state variables such as soil water content or head. The estimation error depends on the measurement error of the state variable and on the uncertainty of the functional relationship. Errors due to bias are in general less significant when the data can be evaluated in terms of change instead of absolute magnitudes.

If an indirectly determined variable depends very sensitively on the actually measured variable as in the case of the relationship between hydraulic conductivity and soil water content, sensitivity analysis is highly recommended. The analysis should not be formal. It should be aimed at the actual problem.

6. Scales

If geostatistics is to be used then there are both theoretical and practical aspects of the relationship between the "support volume" of the samples and the intersampling distances as well as the size of the region of interest. Therefore, the scale of measurements must be considered. In some cases, especially for hydraulic characteristics, there may be an implicit scale in the definition of the characteristic even though this scale may not be well-defined.
Where there are cyclic patterns, the scale (or nested structure of scales) is defined by the wave length(s) and can be determined by means of spectral analysis.

7. Sampling Schemes

Optimal sampling schemes may be developed using the fictitious point method coupled with kriging. With this method, locations for additional measurements are determined in such a way that the measurements will result in a maximum reduction of estimation variance based on the available knowledge.

Field-Scale Water and
Solute Flux in Soils
Monte Verità

EVALUATING THE ROLE OF PREFERENTIAL FLOW ON SOLUTE TRANSPORT THROUGH UNSATURATED FIELD SOILS

Discussion Leaders: W. Jury and K. Roth

Participants: H. Behrendt, M. Cislerová, B. Clothier, G. Destouni, M. Flury, R. Horton, M. Huber, K. Loague, R. Luxmoore, J. Mani, A. Papritz, S. van der Zee, A. Warrick, I. White, A. Zsolnay

Our group discussions on "preferential flow in unsaturated soil" dealt with a number of issues believed to be important in the scientific quest for garnering understanding of this ubiquitous phenomenon. The general areas of discussion were flow mechanisms and classification, modes of appearance, methods of observation, understanding of causal mechanisms, experimental evaluation, and the status of modeling efforts.

Preferential flow refers to the rapid transport of water and solutes through some small portion of the soil volume which is receiving input over its entire inlet boundary. The mechanisms which contribute to preferential flow could include movement through structural voids, unstable flow of the invading fluid, or lateral convergence of water into channels by partial surface clogging or subsurface lateral flow.

1. Modes of Appearance

The group is in complete agreement that preferential flow is widespread. It need not be restricted to situations where the soil architecture has an apparent macrostructure. Indeed, in some macroporous soils, the saturated hydraulic conductivity may be so high that no water application rate is sufficient to fill the macropores. Thus, full Poiseuille flow down the cracks may not occur. Experimental evidence obtained by using dyes, and either adsorbing or anionic tracers have demonstrated that preferential flow may be significant in weakly-structured sandy soils as well.

An obvious relationship is to be expected between soil structure and macropore flow. Soils high in clay that develop cracks will experience preferential flow in these voids when the surface conditions allow them to be filled. Conversely, soils that are coarse-textured and apparently structureless tend to experience wetting-front instabilities more than finer-textured soils, because their permeability has a far greater dependence upon the moisture content.

Although it is not possible with our present state of understanding to predict the prevalence of preferential flow in a particular soil from observations or measurements made in that soil, the physical factors governing preferential flow mechanisms are reasonably well understood. These include the following.

- Rapid flow through filled and partially-filled channels, holes, etc. This phenomenon is restricted to situations where water has a high enough pressure to be drawn into the orifice. Thus, the local soil at the entry point must be sufficiently wet, which is conditional on the local surface boundary condition. In addition, incipient ponding on the less permeable parts of the matrix can access, but not necessarily fill, vented macropores.
- Unstable flow of the invading fluids, initiated by a variety of mechanisms, such as density or viscosity differences, encounter of a subsurface coarser soil layer, or the presence of entrapped air. Other mechanisms thought to induce instabilities are hydrophobic effects in the soil, which are especially common near the surface and alter normal capillary influences on infiltration, and also change in the pressure gradient behind a wetting front, which can trigger an instability at the front.
- Subsurface focusing of water flux into local channels by discrete barriers, such as clay or gravel lenses, or by plant roots.
- Surface focusing of water flux caused by local differences in infiltration rate, perhaps induced by the instantaneous surface boundary condition, or by stem and base flow from plant cover.

Although the understanding of these cause-effect relations is not yet sufficient to predict the extent to which they will occur in given soil, experience and intuition allow us to classify possible behaviour in a preliminary and sensible way. Such a classification may be very useful in developing interim management strategies.

2. Methods of Observation

Members of the group have had considerable experimental experience in the study of preferential flow. It was generally agreed that traditional soil solution samplers are unreliable monitors of preferential flow. Even if the fluid does come in contact with the soil solution extraction volume, it would be unlikely that its passage would occur during the extraction time.

The most reliable methods of observing preferential flow mentioned were dye tracers, soil coring, and soil column effluent experiments in the laboratory. Each has its own serious idiosyncratic limitations.

Soil coring provides a single time- but low volume picture of the resident concentration. However, core samples must be taken at a very high spatial and temporal density to resolve preferential flow patterns. Soil coring does not permit the time scale of preferential flow to be measured, because the preferential flow pattern observed may have ceased moving long before the sampling occurred.

Dyes produce a visual trace of the fluid flow paths, provided that one exposes a trench face after addition of the dye. The price paid for this information is a large hole, and no possibility of observing further movement in that region. Dyes leave a large binary record that does not not allow easy characterization of concentration differences within the stained pattern. Also, the leading edge may become too dilute to be located precisely. Concern was expressed by some members over the poorly-understood chemical properties of many dyes. They might interact with a co-tracer , such as a pesticide, added at the same time, or they might even alter the fluid properties. Soluble dyes such as the anionic amine red may give a misleading picture of the preferential flow domain because of lateral diffusion. More-strongly adsorbing tracers may be useful in tracing structurally-induced preferential flow paths, but they may outline a significantly-different path than the water in a soil experiencing unstable flow. In such cases, the dye should be characterized for adsorptive properties. Some optimism was expressed that statistical characterization of the dye patterns could be performed, thereby providing an easy means by which preferential flow phenomena could be tested for reproducibility, time invariance, etc.

Soil column experiments in undisturbed soils have been used to estimate the speed and character of preferential solute flow in structured soils. The cores must be taken without compressing the interior soil, and the influence of the side walls must be eliminated. General concern was expressed over the manipulation of both the upper and the lower boundary in experiments on such columns. In experiments involving either ponding or excessive irrigation of the inlet end of intact cores, preferential flow velocities of many centimeters per minute were reported, and in extreme cases even faster. However, the laboratory procedure may either enhance or suppress the efficacy of large flow channels. Cores of finite length may create an exit orifice that may not be present in the field. Conversely, the core walls may terminate non-vertical flow channels.

Recent research advances with tension infiltrometers have provided opportunities, in the field, for isolating matrix flow from macropore flow at the infiltration surface, and also at horizon boundaries. However, one of the participants described an unpublished experiment in which methylene blue dye was added with the infiltrating water in a tension infiltrometer, and produced infiltration into cracks which should have been avoided by the device. We did not resolve this apparent conflict with our understanding. Discussions of this experiment raised some interesting speculations about possible effects dyes might have on the contact angle or surface tension of water, as well as some healthy scepticism about whether the infiltrometer in question was maintaining a suction over the surface during the water entry process.

Other methods of observation were discussed briefly. These included the placing of suction barriers laterally into the soil, using large intact monolith lysimeters to study outflow, and making large scale observations of preferential flow by frequent monitoring of tile drain effluents.

3. Understanding of Mechanisms

A strong distinction was made between our understanding of the physics of preferential flow, which was believed to be quite good, and our understanding of the relationship of preferential flow to observable soil features, which was believed to be quite poor. In particular, it would be possible to calculate the extent of flow occurring through structural voids, provided that one knew the surface geometry and water boundary conditions precisely. But, neither of these are observable in soil directly, particularly if the void is below the surface.

Wetting-front instabilities have been studied both experimentally and theoretically. Fingering can arise in a variety of circumstances. An invading fluid can develop instabilities, even in homogeneous soil, if the density or viscosity differ sufficiently from the resident fluid. When heterogeneities are present in the soil, infiltrating water can become unstable even when the soil is relatively dry. A classic case is infiltration of water through fine textured soil overlying a coarser-textured one. Entrapped air and hydrophobicity were cited as two other conditions known to promote instabilities. Hydrophobicity is probably more common than is currently recognized, especially in surface soil, high in organic matter, that has dried out.

Particular importance was ascribed to the condition of the soil surface in inducing preferential flow, both by rendering the entering fluid unstable, and in flow entering structural voids. Instantaneous matrix ponding can generate lateral flow on the microscale. So, a relatively uniform application of water by sprinkler or rainfall can be transformed into a highly nonuniform water entry. This might even produce high enough water potentials to permit water to enter certain channels in structured soils that would be bypassed at the lower area-averaged rate. Thus, it is important to discriminate between the hydraulic characteristics of the matrix, and those of the structural voids. In a coarser textured soil, lateral flow caused by local runoff might produce local regions of high water flux which are large enough to cause water to be diverted into preferential paths right at the surface. It is more likely than not that they will persist to great depth.

Subsurface focusing of the water flux by remnant or occupied root channels, by clay or coarse lenses that are sloped, or by other local soil heterogeneities was cited in a conference presentation. The discussion group felt that this could be an important process whereby preferential flow paths are initiated, especially in coarse-textured soils.

The plant canopy will most likely have a significant effect on localizing incident water fluxes. Stemflow fluxes some hundreds of times the area-averaged rate are not unusual. The biological inhabitants of the soil, and Man's management,were cited in innumerable contexts with reference to the creation or extinguishing of preferential flow. Such near-surface alteration of the soil is conspicuous in creating voids in the soil, and therefore can induce a significant time dependence in soil properties.

In all these cases, the cause of the phenomenon is understood. But, we cannot yet predict it in the complex setting of a field soil. Preferential flow depends on localized soil properties, which must be characterized if prediction of the phenomenon is desired. However, *in situ* measurement problems at this small scale are presently too difficult, but some profitable avenues of research are being followed.

This qualitative-quantitative dilemma was not resolved. But, some possible directions were explored. Perhaps the small-scale patterns of preferential flow can be averaged to produce a stable pattern which is suitable for application. Possibly some form of textural analyses and interpretation of the bulk hydraulic property measurements may reveal attributes that could correlate with preferential flow.

4. Experimental Approaches

It was felt that this area of science was more limited by the dearth of experimental observations than by the availability of experimental models. Thus, some new directions for experiments and novel observation techniques are needed. Multiple tracers, such as adsorbing and mobile ones, together with dyes, or macromolecular tracers that avoid the soil matrix, may help to reveal the preferential flow paths in greater detail. Laboratory experiments, if carefully designed, might be useful in developing some better understanding of the effect of water flow rate and surface conditions on preferential flow.

It was felt that the surface area of soil in contact with a fluid undergoing preferential flow, particularly in structural voids, should be studied in greater detail, especially with respect to its physical and chemical properties. A number of experiments involving strongly adsorbing tracers have shown that preferential flow may occur with little or no solute retardation. This observation cannot be rationally interpreted until the nature of the adsorbing surface is characterized.

More critical effort should be directed towards field experiments. Without a significantly-larger data base, even qualitative generalizations about the relationship of preferential flow to soil features will be difficult to produce. Also, experiments will have to produce more small-scale measurements of soil properties in the vicinity of a preferential flow event. Analyses of the soil physical properties inside and outside of dyed areas may be helpful. Perhaps efforts should be intensified to develop a suction device, such as a large plate or network of samplers. Such an instrument would need to have a high spatial resolution, high flow rate, and rapid response.

5. Modeling Issues

Quantitative modeling of preferential flow processes is in its infancy. Qualitative modeling and rational intuition have long been used to discuss the influence of preferential flow on solute transport. Simple computer and analytical models exist which describe flow of water and solute in two regions of water flow. This is the so-called mobile-immobile or fast and slow velocity representation. Instabilities have also been predicted theoretically using various idealized representations of the soil conditions necessary to create them.

An unresolved issue in such process-oriented models includes the representation of the mass exchange rate between the rapidly moving fluid and the slower moving or stagnant fluid of the matrix. This may be particularly important if the Reynolds number of the preferential flow is high enough to invalidate the Darcy flow approximation. This issue will be critically related to the contact surface area and its properties. Some research is under way in various parts of the world looking at non-Darcy flow in soil, but it is not a well-understood phenomenon.

Concern was also expressed over the types of information required to provide responses to environmental regulatory issues. The arrival time of the center of mass of a pulse or front, or even the first arrival time of a local concentration may be important when preferential flow is involved.

Rate parameters governing the exchange of mass between the preferential flow regions and the surrounding soil matrix are an essential component of all physically-based process models. Yet, they are unobservable experimental quantities, except indirectly through the aid of a specific model of a controlled process. It is also not clear whether effective large-scale parameters embodying the essential characteristics of mass exchange can be developed.

No one in the group felt comfortable proposing specific modeling directions without embarking upon more experimentation. There was general agreement that more comprehensive experiments are needed. A variety of measurements need to be taken, including very small-scale measurements of soil properties, accurate measurements of water flux in preferential flow channels, and a large enough number of solute concentration measurements to assess the extent of preferential flow, its spatial structure, and the location of its point of initiation and termination. Of critical importance is data that will allow assessment of the time invariance of the phenomenon.

There was agreement that existing understanding of preferential flow allows rational management decisions to be made, even if the soils that exhibit it cannot be modeled at the field scale. Our understanding of cause-effect relationships obtained in controlled settings in the laboratory and the field may allow us to generate the understanding required to manage natural soil in a logical way.

Many of the issues raised were not resolved in discussion. Preferential flow is a generic term referring to a host of processes that may require specific experimental or theoretical approaches. At the end of the sessions it was concluded unanimously that the discipline would benefit from a stronger classification system, perhaps relating preferential flow to a stability index or to some fluid property like Reynolds number.

Part 2

Papers

Field-Scale Water and
Solute Flux in Soils
Monte Verità

SPATIAL VARIABILITY OF WATER AND SOLUTE FLUX IN A LAYERED SOIL

R.G. Kachanoski, C. Hamlin, and I.J. van Wesenbeeck

All soils have horizons (layers) but their influence on water and solute flux have not been studied in any detail. Solute transport parameters were examined in a sandy soil by applying a pulse of KCl to the soil surface and measuring breakthrough curves in both the Ap horizon and B horizon using solution samplers. Measurements were taken under steady surface flux density of water applied using drip lines. The solute velocity in the B horizon was significantly faster than the A horizon and was directly related to a lower transport volume in the B horizon. Solute velocity in the B horizon could not be accurately estimated without knowing the thickness of the Ap horizon. The measurements suggested a transfer function model (correlated flow) would accurately predict the variance of solute travel times, if the variance prediction was corrected for faster mean solute travel times in the B horizon. The study indicates the importance of obtaining both solute transport and soil survey information at a site.

1. Introduction

The purpose of any spatial model is to simplify, organize, and extrapolate information about a system. Soil systems are particularly complex because of the large number of interacting variables. Water and solute flux at a particular spatial scale of interest will inherently reflect the variability of controlling soil properties. Thus, an understanding of the influence of soil properties on transport is important in determining the accuracy in extrapolation (prediction) of any spatial transport model. Most water and solute flux studies identify a specific soil series for the study site. This by definition connects the water and solute flux behaviour to a soil classification system.

Soil classification systems are usually based on the concept of a pedon which is defined as the smallest three dimensional spatial unit of the surface of the earth that is considered as a soil (Soil Survey Staff, 1960). The three dimensional spatial unit (pedon) is further subdivided into soil horizons, which are characteristic soil layers formed by the chemical, physical, and biological transformation of the original parent material (porous, non-soil material). In most soils, horizon thicknesses are <1 m. It is the presence (or absence) of horizons and combination of horizons, along with their average (within defined limits) soil properties, which place an individual spatial soil unit into a classification system. Significant properties used to characterize horizons are texture (% sand, silt, clay), organic carbon (%), pH, structure (organization of solid material and voids), and horizon thickness. All of these properties have been shown to influence either water or solute flux.

Since all soils have at least one horizon, which is different than the underlying parent material, validation of the physics of a solute transport model that assumes a homogeneous porous medium is a moot point. Little information is available on the influence of soil horizon boundaries on solute transport characteristics.

The objective of this paper is to examine insitu the influence of an A/B horizon boundary on solute transport characteristics.

2. Solute Transport Models

Solute transport in the unsaturated zone at the field scale has been modelled using three approaches: (1) the convective dispersion equation (CDE) with constant coefficients, (2) the CDE with spatially distributed coefficients, and (3) the transfer function method (TFM) or equivalent stochastic convective method which uses solute travel time probability distribution functions (pdf's). The three approaches are the subject of a number of the papers in these proceedings and have recently been reviewed (Butters and Jury, 1989).

All three modelling approaches predict the same average solute transport velocity, but differ in their predicted "spread" or dispersion of a solute pulse moving through the soil. A simplistic but useful comparison of the CDE and TFM can be given in terms of the occurrence (or non-occurrence) of correlated travel times with depth (Jury, 1982).

The CDE assumes the solute transport velocity is the same at every location, both in the horizontal and vertical spatial planes. Differences in arrival time at some depth $Z = L$ of a conservative solute are assumed to occur by a random diffusion/dispersion process. A solute particle displaced in the opposite direction relative to the convective flow at one depth could just as easily be displaced forward in the next depth increment. Thus, the travel time of a solute particle to depth $Z = L$ is assumed to be uncorrelated to its travel time in the next depth increment.

In contrast, the TFM assumes that at any particular location the solute velocity is constant with depth, but this velocity varies in the horizontal plane. Solute "spreading" of a pulse at the field scale is attributed to the horizontal spatial variability in vertical transport velocities. Since the vertical solute velocity at any particular location is constant with depth, the horizontal spatial pattern of solute travel time to depth $Z = L$ is correlated to the spatial pattern of travel times in the next depth increment.

The total travel time $t_{2L,i}$ of a solute particle moving through two layers (L_1, L_2) of equal thickness at location $i = (x,y)$ in the horizontal plane can be given by

$$t_{2L,i} = t_{L1,i} + t_{L2,i} \tag{1}$$

where $t_{L1,i}$ and $t_{L2,i}$ are the travel times to pass through layers L_1 and L_2 respectively. The mean or expected travel time across all locations in the horizontal plane is then given by

$$E(t_{2L}) = E(t_{L1}) + E(t_{L2}) \tag{2}$$

where $E(\surd)$ is the expectation operator. For both the TFM and CDE, the $E[t_{L1}] = E[t_{L2}]$ and the average travel time through two depth increments of equal thickness is equal to two times the travel time through one of the depth increments.

The "spread" of a solute pulse can be characterized by the variance of the solute travel time pdf. Applying the variance operator $Var(\surd)$ to equation (1) gives

$$Var(t_{2L}) = Var(t_{L1}) + Var(t_{L2}) + 2\,Cov(t_{L1}, t_{L2}) \tag{3}$$

where Cov (t_{L1}, t_{L2}) is the covariance between the travel times of a solute particle in the two depth increments.

For the CDE (uncorrelated flow) the $Cov(t_{L1}, t_{L2}) = O$ and Var $(t_{L1}) = Var(t_{L2})$, thus

$$Var(t_{2L}) = 2 \cdot Var(t_{L1}) \tag{4}$$

Since L_1 can represent any thickness, it can be easily shown that the CDE (uncorrelated flow) predicts that the solute time travel variance increases linearly with depth.

For the TFM (correlated flow) $t_{L1,i} = t_{L2,i}$. Thus,

$$2\,Cov(t_{L1}, t_{L2}) = 2\,Var(t_{L1}) \tag{5}$$

and Var (t_{L2}) = Var (t_{L1}), which substituted into equation (3) gives,

$$Var(t_{L2}) = 4 \cdot Var(t_{L1}) \tag{6}$$

Again since L_1 can represent any thickness it can be shown that the TFM (correlated flow) predicts that the solute travel time variance increases as the square of distance. This is equivalent to having a dispersion coefficient which grows linearly with distance (Jury and Sposito, 1985).

The difference between the TFM and the CDE with respect to solute spreading can be viewed as the presence or absence of the covariance term between travel times through successive depth increments. The above analysis holds only for soils where $E[t_{L1}] = E[t_{L2}]$.

3. Materials and Methods

The study site was located at the Delhi Research Station in Southwestern Ontario, Canada. The soil is classified as a Fox sand and has an Ap horizon with an average thickness of 0.25 m (Table I). The field saturated hydraulic conductivity K_{fs} and the inverse flow weighted macroscopic capillary length scale α (White and Sully, 1987) for the horizons were measured in the study area using a pressure infiltrometer (Reynolds and Elrick, 1989). The K_{fs} is significantly higher in the B horizon compared to Ap horizon (Table I). The average α value is also larger in the B horizon. The hydraulic measurements indicate the B horizon is a layer with a greater proportion of large pores which transmit water faster than the overlying Ap horizon.

Table I. Average soil properties for study site.

Horizon	Thickness m	K_{fs}[1] $m\ s^{-1}$	α[2] m^{-1}
Ap	0.25	4.0x10-5	9.6
Bm	0.80	16.3x10-5	12.3

[1]K_{fs} = field saturated hydraulic conductivity
[2]α = inverse flow weighted macroscopic capillary length

Forty-eight (N=48) soil solution samplers with porous ceramic cups were installed along a transect at 0.2 m intervals and 0.2 m depth. A second set of solution samplers (N=48) were also installed adjacent to the first samplers, but at a depth of 0.4 m. The 0.4 m samplers were installed at an angle to place the sampling cups almost directly below the 0.2 m sampling cups. All samplers were installed by hand auguring with a special auger constructed with a drill bit with an outer diameter equal to the ceramic cup diameter (10 mm). The site was free of any vegetation. All of the samplers were connected to a single vacuum system so that solution samples could be collected from all locations simultaneously.

Soil water content was measured using the time domain reflectometry TDR method using thin (2mm diam.) stainless steel rods as transmission lines, and a Tectronix 1502b cable tester to measure dielectric constants. Transmission lines 0.4 m deep were installed beside every 0.4 m deep solution samplers. Soil water contents were calculated based on the calibration curve of Topp et al (1980). Measurements were taken 3-4 times daily during the initial wetting up of the soil profile, and before and after the pulse experiment.

A constant flux density of water was applied to an area (2m x 12m) surrounding the instrumented transect. Water was applied using drip lines with emitters spaced every 0.15m along the lines. Adjacent lines were spaced 0.1 m apart. Water was applied at a rate of q= 35 mm hr^{-1} for at least 48 hours until the soil water content measurements indicated the upper soil profile was at steady state (ie. $\partial\theta/\partial t = 0$). At this time (t=0), a pulse of KCl was applied to the soil surface, equivalent to 100 g Cl m^{-2}. The pulse was added using hand sprayers containing KCl dissolved in water. Approximately 1 L m^{-2} of concentrated KCl solution was applied. Soil solution samples were subsequently taken at the 0.40 depth every 5-10 minutes over a total time of approximately 600 minutes. A second pulse of chloride was subsequently added and the sampling repeated for the 0.20 m depth. Separate pulses were added to eliminate any influence of sampling at the 0.2 m depth on measurements at the 0.4 m depth. Soil solution samples were analyzed for chloride on a Traacs 800 autoanalyzer.

The thickness of the Ap horizon at each location in the transect was determined by auguring at the end of the experiment.

Average breakthrough curves $\overline{C}_L(t)$ were calculated for the 0.2 m and 0.4 m depth by calculating the average concentration of Cl across all samplers (N=48), at each sampling time. Since the Cl was added as a narrow pulse the solute travel time probability density function $f_L(t)$ for L=0.2 m and 0.4 m was estimated from the breakthrough curves (Jury et al. 1982; Simmons, 1982; White et al. 1986)

$$f_L(t) = \overline{C}_L(t) / \int_0^T \overline{C}_L(t)\, \partial t \tag{7}$$

where T= total time of measurement. The $f_L(t)$ can be expressed as a function of total cumulative infiltration I (ie. $f_L(I)$) using the identify I = qt.

The total mass recovery of chloride per unit area M_L at each depth was calculated from

$$M_L = q \int_0^T \overline{C}_L(t)\, \partial t \tag{8}$$

Moment analysis was conducted on the measured $f_L(t)$ to determine the effects of crossing the horizon boundary on solute dispersion. The average or expected solute travel time $E_L[t]$ to depth L was estimated from the first moment,

$$E_L(t) = \int_0^T f_L(t)\, t\, \partial t \tag{9}$$

The variance of the solute travel time probability density function $Var_L(t)$ was subsequently calculated

$$Var_L(t) = \int_0^T f_L(t) \left[t - E_L(t) \right]^2 \partial t \tag{10}$$

The estimates $f_L(t)$, $E_L(t)$, and $Var_L(t)$ are associated with the scale of the entire transect. Similar estimates of these parameters can be obtained for each location by normalizing individual BTC's. In a manner similar to Butters and Jury (1989), the average of the individual $f_L(t)$, $E_L(t)$, and $VAR_L(t)$ will be called local (near-field) estimates while the values estimated from the transect averaged BTC $C\overline{C}_L(t)$ will be referred to as field scale (far-field) estimates. The far-field estimate of $Var_L(t)$ includes solute velocity variations over the entire transect while the near field estimate is an average of the local scale variations (single solution sampler) of solute velocity.

4. Results and Discussion

The field scale BTC's $\overline{C}_L(t)$ for the 0.2 m and 0.4 m depths are given in Figure 1. Total mass recovery estimated with equation (8) varied from 20% to 190% of applied for individual BTC's and averaged approximately 70%. The average solute travel time at the field scale was 1.78 hr

and 3.28 hr for the 0.2 m and 0.4 depths respectively (Table II). The average travel times indicate that the solute velocity V increased from 0.112 m hr^{-1} in the first 0.2 m, to 0.133 m hr^{-1} in the 0.2 m to 0.4 m layer. The increase in velocity in the 0.2 m to 0.4 m layer is consistent with the significantly higher hydraulic conductivity in the B horizon. However, the 0.133 m hr^{-1} velocity in the 0.2 m to 0.4 m layer is not the velocity in the B horizon, since the average Ap horizon thickness was 0.25 m. The first set of solution samplers was placed at 0.2 m to ensure that all of them were within the Ap horizon, which varied from 0.2 m - 0.3 m in thickness.
The average travel time in the B horizon was calculated by assuming the pulse travelled at 0.122 m hr^{-1} for 0.25 m. This gave an estimated average solute travel time in the Ap horizon of 2.33 hr. The average travel time in the 0.25 m to 0.40 m of the B horizon was then estimated at (3.28 - 2.23 hr) = 1.05 hr. This gave an average B horizon velocity, V_B = 0.142 m hr^{-1}. Separate experiments (not given) with solution samplers at 0.4 m and 0.8 m depths have indicated the average solute velocity in the 0.4 m to 0.8 m depth increment was 0.140 m hr^{-1}, which is similar to the estimated B horizon velocity after correcting for the Ap horizon thickness.

Table II. Average solute transport properties.

	Local scale		Field scale	
Depth [m]	E(t) [hr]	Var(t) [hr^2]	E(t) [hr]	Var(t) [hr^2]
0.2 m	1.68	0.21	1.78	0.30
0.4 m	3.10	0.78	3.28	1.04

The average water filled pore volume $\bar{\theta}$ (m^3m^{-3}) participating in solute transport can be estimated from

$$\bar{\theta} = q/V \tag{11}$$

which gives $\bar{\theta}_A = 0.31$ m^3m^{-3}, and $\bar{\theta}_B = 0.25$ m^3m^{-3} for the A and B horizons respectively. The average transport volume for the 0-0.4 m depth is estimated at 0.287 m^3 m^{-3} which is very similar to the measured TDR soil water content of 0.26 m^3 m^{-3}.

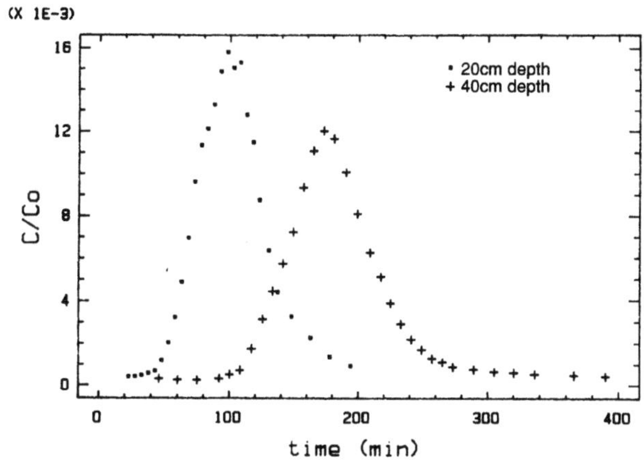

Figure 1. Field scale solute breakthrough curves.

The solute travel time variances at the field scale were 44% and 33.0% higher than the local scale variances for the 0.2 m and 0.4 m depths respectively (Table II). This is similar to observations reported by Butters and Jury (1989). The increase in variance at the field scale is related to the scale dependence of the lateral variations in solute velocity. The space-time solute concentration contours for the 0.4 m depth indicated significant differences in BTC's across the measurement transect (Fig. 2). The average solute travel times tend to be longer (slower average velocity) near the end of the transect. This is attributed to an increase in Ap horizon thickness for that portion of the transect (Fig. 3). As shown earlier, the transport velocity in the Ap horizon is slower than the B horizon. Thus, locations which have a thicker Ap horizon will have a longer average solute travel time to the 0.4 m depth. The 50 μg/g time contour line from Figure 2 (i.e. the time it took for the BTC to come back down to 50 μg/g) has a spatial pattern very similar to the pattern of the Ap horizon thickness (Fig. 3). This indicates that local fluctuations in Ap thickness dramatically affected the local BTC and thus would be responsible for at least part of the increased variance at the field scale.

The Var(t) of the 0.40 m depth was 3.7x and 3.5x the Var(t) estimate of the 0.20 m depth for the local scale and field scale respectively. This is not quite a 4x increase in variance as predicted for correlated flow (TFM) but it is significantly higher than the 2x increase predicted by uncorrelated flow (CDE). However, the 4x increase (equation (6)) is for the case of constant velocity with depth which is clearly not the case in this soil (Table II). Assuming that the solute travel time in the upper 0.2 m was different than, but still spatially correlated to the travel time in the second layer (0.2 - 0.4 m), then it follows that

$$\mathrm{Var}_{0.4}(t) = \left[E_{0.4}(t)/E_{0.2}(t)\right]^2 \mathrm{Var}_{0.2}(t) \tag{12}$$

where the subscripts on the statistical operators indicate the depth (m) of measurement. Using the mean travel times in Table I, equation (12) predicts that the $\mathrm{Var}_{0.4}(t)$ will be 3.4x the $\mathrm{Var}_{0.42}(t)$, if correlated flow occurred across the horizon boundary. This is very similar to the measured change in travel time variance. The similarity in the measured change in Var(t) with depth and the predicted change for correlated flow does not constitute proof of the existence of correlated flow. The dispersion coefficient (CDE) of the B horizon may be larger than the Ap horizon, and this may also explain the data. An additional pulse experiment on the same transect was carried out after the Ap horizon was removed. The samples from this experiment are currently being analyzed and will give an estimation of the B horizon transport characteristics without the influence of the A horizon.

5. Summary

This field study indicated the average solute travel times at the 0.2 m depth could not be used to predict the travel times measured at greater depths. The B horizon had a significantly lower calculated transport volume and measured steady state water content, which caused the solute velocity to increase. A second measurement of the average solute travel time at 0.4 m also would not predict the average travel times at greater depths unless the average thickness of the Ap horizon was known and a correction calculated.

The variance of the solute travel times increased by a factor 3.5x with a doubling of the transport depth. This result indicated that correlated flow may be occurring across the horizon boundary and the TFM approach to predicting solute transport may be best for this situation. If the TFM model is used, then the estimated solute travel time variance will also have to be corrected to account for the change in velocities in the horizons. This correction depends on the average horizon thickness relative to the prediction depth. If the CDE is used a different dispersion coefficient will be needed for each horizon and the average horizon thickness will be needed for subsequent prediction.

The study points out the importance of measuring both solute transport and soil morphological data (i.e. horizon properties).

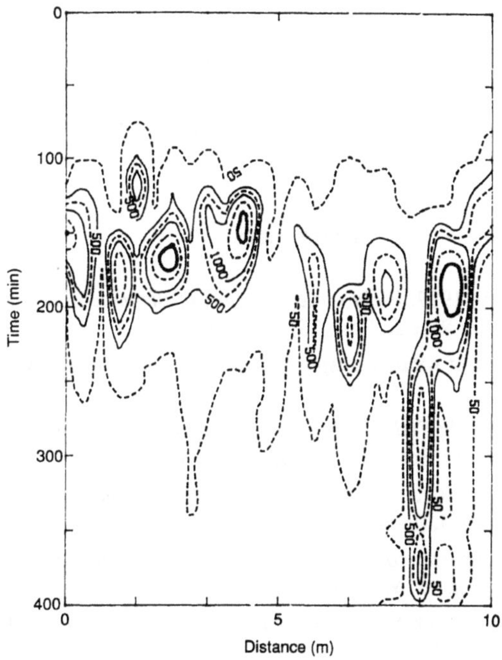

Figure 2. Solute concentration (μg/g) contours in space and time for the 0.4 m depth.

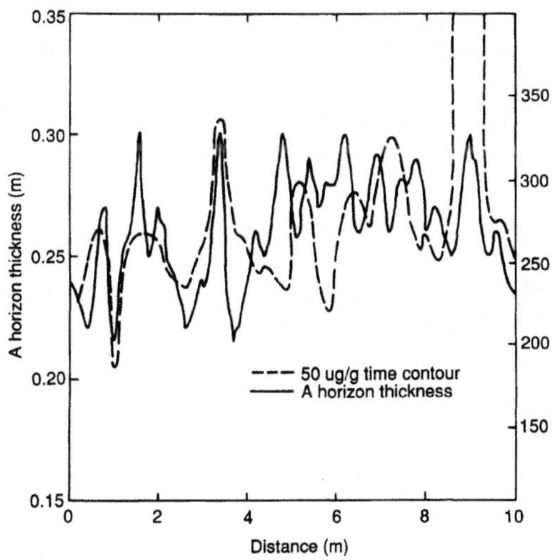

Figure 3. Spatial pattern of the Ap horizon thickness and the time for the local breakthrough curves to come back down to 50 μg/g.

Acknowledgements

This work was supported by the Natural Sciences and Engineering Research Council (NSERC) of Canada, and Agriculture Canada.

References

Butters, G.L., and W.A. Jury. 1989: *Field scale transport of Bromide in an unsaturated soil. 2. Dispersion modeling*. Water Resour. Res., **25**, 1583-1589.

Jury, W.A., 1982: *Simulation of solute transport using a transfer function*. Water Resour. Res., **38**, 363-368.

Jury, W.A. and G. Sposito, 1985: *Field calibration and validation of solute transport models for the unsaturated zone*. Soil Sci. Soc. Amer. J., **49**, 1331-1441.

Reynolds, W.D., and D.E. Elrick, 1989: *Steady infiltration from a ponded ring source: I. Theory*. Soil Sci. Soc. Amer. J. (in press).

Simmons, C.S., 1982: *A stochastic-convective transport representative of dispersion in one-dimensional porous media systems*. Water Resour. Res., **18**, 1193-1214.

Soil Survey Staff. U.S.D.A., 1960: *Soil Classification, 7th approximation*. U.S. Government printing office, Washington, D.C. 265 p.

Topp, G.C., J.L. Davis, and A.P. Annam, 1980: *Electromagnetic determination of soil water content: Measurement of coaxial transmission lines*. Water Resour. Res., **16**, 574-582.

White, R.E., J.S. Dyson, R.A. Haugh, W.A. Jury, and G. Sposito, 1986: *A transfer function model of solute transport through soil, 2. Illustrative applications*. Water Resour. Res., **22**, 248-254.

White, I., and M.J. Sully, 1987: *Macroscopic and microscopic capillary length and time scales from field infiltration*. Water Resourc. Res., **23**, 1514-1522.

R.G. Kachanoski, C. Hamlin, and I.J. van Wesenbeeck, Dept. of Land Resource Science, University of Guelph, Guelph, Ontario, Canada N1G 2W1

Field-Scale Water and
Solute Flux in Soils
Monte Verità

ONE AND THREE DIMENSIONAL EVALUATION OF SOLUTE MACRODISPERSION IN AN UNSATURATED SANDY SOIL

G. L. Butters, T. R. Ellsworth, W. A. Jury

The results from a series of one-dimensional and three-dimensional solute transport studies done at the University of California, Riverside, are reviewed and integrated. It is found that the stochastic-convective lognormal transfer function model (CLT) after a single calibration at 0.3 m provides an excellent simulation of the observed field scale breakthrough curves to a depth of 1.8 m under both transient and steady-state water flow conditions. By 14 m travel distance, the dispersivity appears to be converging to an asymptote for small areas (~3 m^2) within the field but the area-averaged macrodispersivity for the entire 640 m^2 field continues to grow at this distance. The effect of soil texture changes in this field is to produce oscillations in the longitudinal variance of a solute plume without appreciably increasing transverse dispersion. A method of linking the parameters of the CLT model to changes in soil properties is tested and found to have the potential to improve the original CLT model predictions of solute dispersion beyond the region of variable soil texture.

INTRODUCTION

The lack of comprehensive field studies of solute movement is often cited as a major impediment to our understanding of solute transport in heterogeneous systems. Rare is the opportunity to analyze detailed complementary data sets from the same field, since the process of observation at this scale is expensive and often destructive. Recently, however, research at the University of California, Riverside, has generated comprehensive data sets from a series of large scale, one and three dimensional solute transport studies conducted over an eight year period. As much has already been written regarding the first of these studies (Butters, 1987; Butters and Jury, 1988, 1989; Butters et al., 1989a,b), this paper is an attempt to integrate the findings of all of the experiments into a consistent predictive model of solute movement at the field site.

As a prerequisite to discussion of the solute experiment, we shall begin by describing the physical properties of the field site. A summary of each of the experimental procedures follows, illustrating the sequential nature of our understanding of the solute dispersion process and attempts to model it.

Physical Setting

The 80 m x 80 m field site located near Riverside, California, consists of a young, well drained, alluvial loamy sand classified as a mixed, thermic, Typic Xeropsamments by the USDA taxonomic system. A detailed description of the soil texture and bulk density (Bd) characteristics as a function of depth is shown in Figure 1. The Bd profile (Ellsworth, 1989; Ellsworth et al., 1989) represents the average of three pits where the samples were taken from the sidewall at 0.1 m intervals to a depth of 5 m. Included in Figure 1 is the volumetric water content profile (estimated as the product of bulk density and gravimetric water content measurements) during the one-dimensional steady state flow experiment to be discussed shortly. The soil is fairly uniform over the upper 2.5 m, with a zone of finer texture and reduced Bd in the vicinity of 3 m. The soil structure in these regions is single-grained (structureless) and subangular blocky, respectively. The water content profile reflects these changes with a maximum in the low Bd, fine-textured region and a minimum in the high Bd, coarser-textured zones. Not evident in the data is the sporadic occurrence of coarse sand and gravel lenses near a depth of 0.9 m, which were observed in several studies at the site. The soil below 5 m to a depth of 25 m is a strata of sands and loams (data not shown). The saturated zone beneath the field site is reported to be in excess of 150 m from the surface.

In addition to the static soil properties reported above, water flow properties of the field site have been assessed using a variety of techniques (unpublished data). Using an air-entry permeameter (AEP)(Topp and Binns, 1976) at 54 locations the saturated hydraulic conductivity (K_S) at the soil surface was found to be normally distributed with a mean of 0.062 m/hr and a coefficient of variation CV of 0.44. The variogram of the AEP-K_S was described best using a spherical model and had an integral scale of 7.1 m.

Subsurface K_S measurements were accomplished using a Guelph permeameter (GP) (Reynolds and Elrick, 1985) at seventy-four closely spaced locations. The GP-K_S at 0.3 m was lognormally distributed with a mean of 0.26 m/hr and a CV of 0.5. The variogram of GP-K_S was described best using an exponential model and had an integral scale of 3.4 m. While direct comparison of the AEP-K_S and GP-K_S is not valid, it seems evident that the subsurface flow features have greater lateral variability than the surface measurement implied.

The rapid drainage features of the site are indicated in Figure 2. Using the simplified instantaneous profile method, the figure illustrates the $\theta(z)$ profile between the soil surface and 1.2 m depth at several times following saturation. Notice that much of the drainage observed after 22 days occurred in the first day.

Experimental Summaries and Major Findings

The first one-dimensional solute transport study at the field site was reported by Jury et al. (1982). Sixteen sampling locations separated by 20 m and arranged in a regular 4x4 grid pattern contained porous ceramic solution samplers at depths of 0.3, 0.6, 0.9, 1.2 and 1.8 m. Following the sprinkler application of a narrow pulse of KBr, the Br-concentration of the soil solution at the

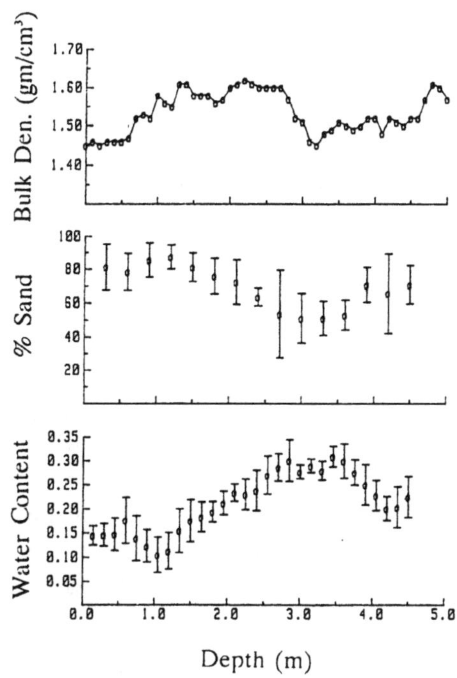

Figure 1. Soil texture (% sand), bulk density, and water content profile to a depth of 5 m. The error bars represent 95% confidence intervals.

five depths was monitored over several months during leaching by erratic rainfall. The results of the study (hereafter called the transient flow experiment) were used to test the stochastic-convective lognormal transfer function model (CLT) proposed by Jury (1982). The test of the CLT was inconclusive. Because of incomplete recovery of the solute plume at depths below 0.9 m, uncertainty in the solute travel time variance at these depths rendered the data set unable to distinguish contradictory models of the solute dispersion process (Jury et al., 1988).

In 1984, the sampling network established for the transient study was expanded by adding sixteen solution samplers at 3.05 m and six samplers at 4.5 m (Butters, 1987; Butters et al., 1989). Using bidaily sprinkler irrigations to establish a quasi-steady state flow regime, a second pulse of bromide was injected and monitored at each of the seven depths for several months (hereafter called the steady-state experiment). This time the bromide recovery per unit area at each depth was nearly 100% of the applied mass. At the conclusion of the experiment, six soil cores were taken in 0.3 m intervals to a depth of 25 m, extracted, and analyzed for Br^- in order to characterize the final Br^- depth profile.

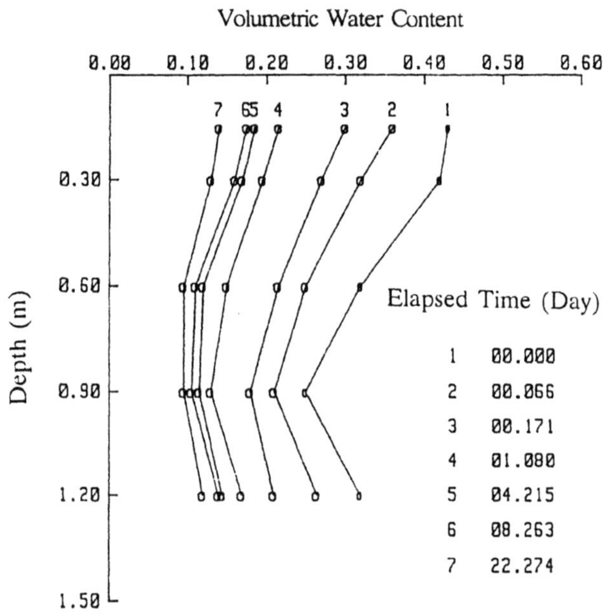

Figure 2. Water content profiles at several times following saturation (Data courtesy of Dr. Shouse, U.S. Salinity Laboratory).

The major finding of the steady-state experiment (Butters and Jury, 1989) was that the variance of the travel time distribution between the surface and 3.05 m increased nearly as the square of the distance traveled and thus a linear scale effect in the longitudinal macrodispersivity was evident. The linear growth of the longitudinal macrodispersivity was predicted by the stochastic-convective CLT model, which assumed that the soil was vertically homogeneous (stationary) and that lateral mixing of solute was insignificant. Figure 3 shows the normalized breakthrough curves (BTC's) from the steady-state experiment (solid line) at four depths plotted as a function of net applied water together with the CLT prediction after calibration at 0.3 m. Also plotted in Figure 3 are the BTC's from the transient experiment. The model prediction is in excellent agreement with the main features of the BTC's under the transient water flow conditions. This remarkable similarity of the BTC's from the two experiments implies that the calibration BTC at 0.3 m may be a fundamental property of this field site and that it may be used to predict area-averaged solute transport accurately through the vertically homogeneous soil (to a depth of 2.5 to 3 m) even under transient water flow conditions. Solute dispersion is apparently dominated by the lateral velocity variation which persist through the upper 3 m and, because the rapid drainage

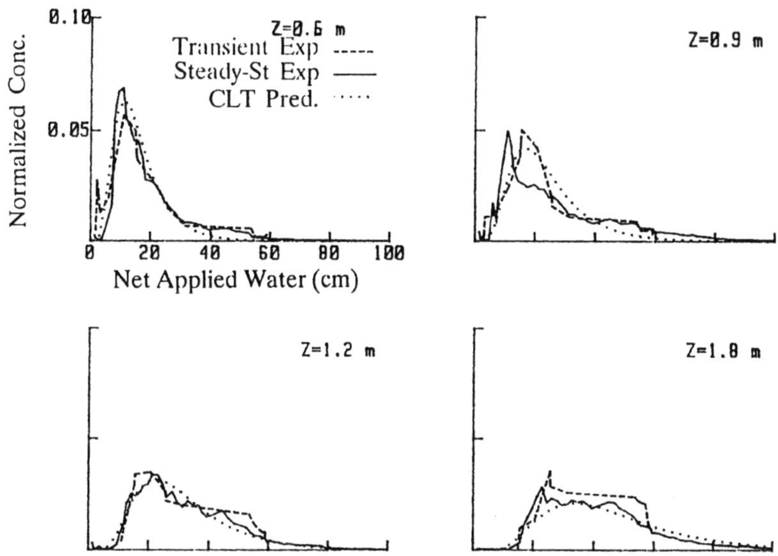

Figure 3. Normalized BTC's from the steady-state (solid line) and transient (dashed line) flow experiments at depths, z, of 0.6m, 0.9 m, 1.2 m, and 1.8 m. The dotted line is the CLT prediction after calibration at 0.3m.

and coarse texture (Figs. 1 and 2), chemical movement shortly after an infiltration event is minor and quickly reversible by subsequent events (i.e. no entrapment in dead-end pores). Also, the rapid drainage of the soil after irrigation allows the use of net applied water as the independent variable to scale chemical transport under different conditions, even though cumulative drainage past the depth of observation is more appropriate (Jury et al., 1989).

While the CLT simulation of the solute dispersion over the upper 3 m was quite good, changes in the soil texture and structure beyond 3 m strained the model's vertical homogeneity assumption. By 4.5 m travel distance the growth rate of the travel time variance and the magnitude of the dispersivity had decreased markedly and the CLT overpredicted the dispersion. The difference between the BTC at 4.5 m (Fig. 4) and the CLT prediction illustrates a change in the longitudinal spreading in the region between 3 and 4.5 m. This change, however, appeared to be short lived, as the dispersivity increased three-fold by the final observation of the pulse with the analysis of the deep soil cores. Figure 5 plots macrodispersivity (field scale) and the local dispersivity (avg. of individual sites) as a function of depth. The dispersivity was calculated from the best-fit parameter estimates of the appropriate solution of the convection-dispersion equation (Butters and Jury, 1989). The increase in the local dispersivity beyond 4.5 m is modest compared to that of the macrodispersivity and may suggest approach to Fickian solute spreading at the local scale

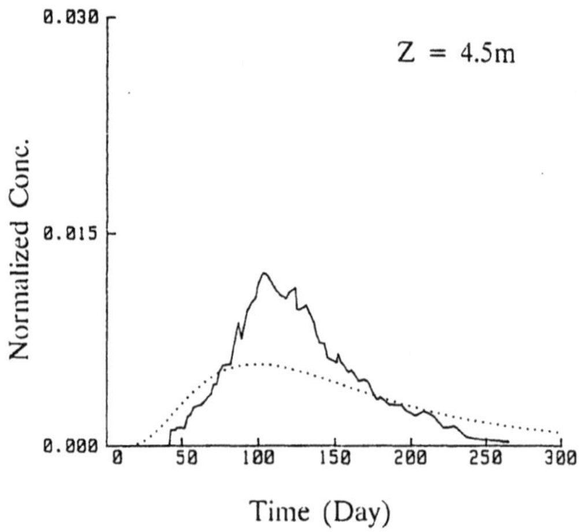

Figure 4. Normalized BTC at 4.5 m depth (solid line) together with CLT prediction after calibration at 0.3 m.

(~3 m^2). The large increase in the macrodispersivity does not suggest convergence but a definitive conclusion is tempered by the uncertainty inherent from the analysis of only six locations in a 640 m^2 area. A drop in the dispersivity between 3.05 m and 4.5 m occurs at both scales and even if only those six sites which have both 3.05 m and 4.5 m samplers are considered (not shown). This indicates that the dispersivity decrease between 3.05 m and 4.5 m is not simply the result of fewer sampling sites at 4.5 m. Another interesting feature in Figure 5 is the elevated macrodispersivity at 0.9 m. We speculate that this results from fingers of gravel present at only a few sites which affected either the chemical movement or the sampler reliability.

At the conclusion of the one-dimensional steady-state experiment, we hypothesized that the decrease in the longitudinal dispersivity between 3 and 4.5 m might be the result of increased lateral mixing (transverse dispersion) in the finer textured soil near 3 m (see Fig. 1). This hypothesis was tested by a third major study (Ellsworth, 1989; Ellsworth et al., 1989) (referred to hereafter as the three-dimensional study) in which the shape of a solute plume was resolved at three times after injection. A chloride solution was applied to each of eight plots (2.25 m^2 to 4.5 m^2) over a two-week period while surrounding areas received an equivalent volume of solute-free sprinkler irrigation. These cubic plumes were then pushed through the soil with solute-free irrigation and their shapes characterized by extracting closely-spaced soil cores (about 500 samples between the surface and 5 m depth for each plot and each sampling time). The solute recovery ranged from 78% to 138%. The solute plumes were observed to compress longitudinally without a significant increase in transverse dispersion as they moved through the fine textured zone near 3 m. Figure 6 illustrates a typical plume two dimensionally at three times after injection (25, 48, and 63 days). In Figure 6a, the center of mass of the plume at the three sampling times is 2.4, 3.6, and 4.3 m, respectively. Observe that despite solute movement into and through the fine textured zone between 3 m and 4 m, there is no noticeable increase in the lateral dimensions of the plume. Figure 6b, shown as if viewed through an excavated sidewall, depicts the vertical extension of the plume at the three sampling times. The vertical dimension of the plume at the three sampling dates was about 3.2, 2.5, and 3.0 m, respectively. Thus, the plume appears o have compressed by the second sampling time and expanded by the third.

Among its major findings, the three-dimensional study provided strong evidence that the drop in the dispersivity observed between depths of 3 and 4.5 m in the one-dimensional experiment was not due to large-scale lateral mixing. After analysis of twelve different plots within the field, Ellsworth concluded that transverse solute spreading of less than 0.5 m could be expected over the entire upper 5m of soil under steady state water flow conditions. Thus the site or local scale in the one-dimensional experiment is only slightly larger than the spatial separation of the solution samplers at a given site. Furthermore, the degree of lateral movement is far less than the integral scale (J_2) of the near surface solute velocity or saturated hydraulic conductivity (J_2 = 2.1 m and 3.4 m, respectively) and indicates that the transverse mixing is insufficient to smooth out the lateral variation in the solute velocity within at least the upper 2.5 m of homogeneous soil. As a result, we observed a dispersivity scale effect in the one-dimensional study and we were able to successfulle employ a purely stochastic-convective model. By 12 m, the solute transit through the soil layers had apprently caused enough interruption to damp out convective extremes within the local scales (i.e. individual sites) and the dispersivity appears to be converging to a constant value (see Fig. 5). However, the differences in mean convection between the different sites persist as the solute moves through the layered material to an extent that asymptotic Fickian behavior is not evident at the field scale.

Modeling

Using the results of the studies discussed above, our attempts to develop a rigorous analytical model to predict solute macrodispersion after passage through the heterogeneous soil zone have thus far been unsatisfying. The CLT model, which treats the soil as though it was composed of a series of isolated stream tubes with different velocities, provided an excellent description of solute spreading over the vertically-homogeneous upper 1.8 m. Upon entry into the fine textured soil layer, however, a disruption in the growth rate of the effective solute macrodispersion coefficient is evident. Because the layer has a higher water retention than the coarser soil above it, the solute velocities are reduced in this zone, principally because the water flux is spread through a larger transport volume. The apparent decrease in the spreading of the solute pulse observed in the three-dimensional study can be explained simply as a consequence of the reduction in solute velocity in this zone, which allows the trailing edge of the plume within the upper soil to catch up to the leading edge within the fine-textured region. However, this argument cannot explain the change in the shape of the one-dimensional travel time distribution observed at a depth below the texture change, since all molecules travel the same distance and are slowed equally.

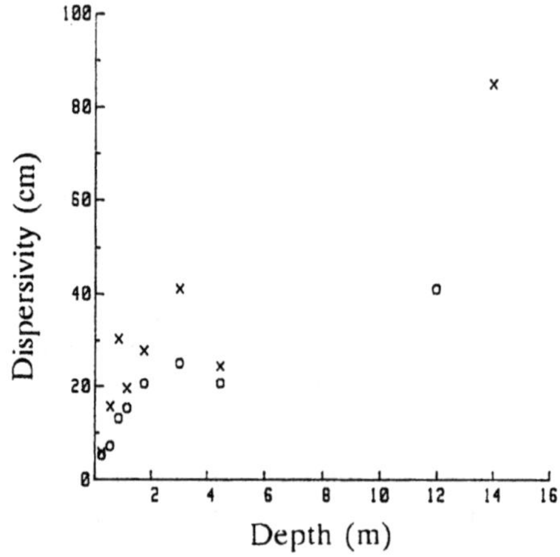

Figure 5. The dispersivity as a function of travel distance determined from the field average BTC (X's) and values from individual BTC's within the field (O's).

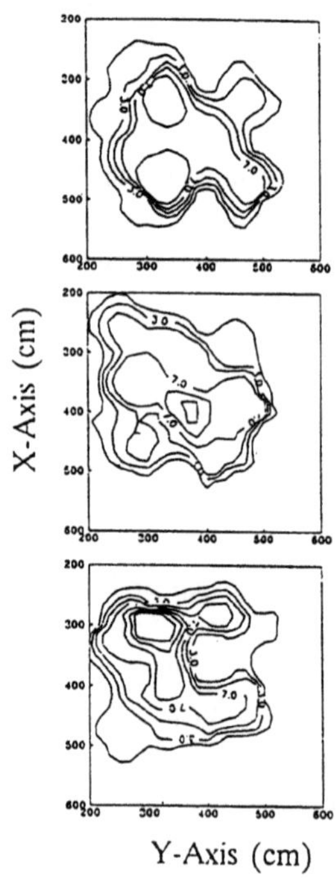

Figure 6a. An overhead two-dimensional view of the solute plume in the 3D-study at three times after injection (top 25, middle 48, and bottom 63 days)

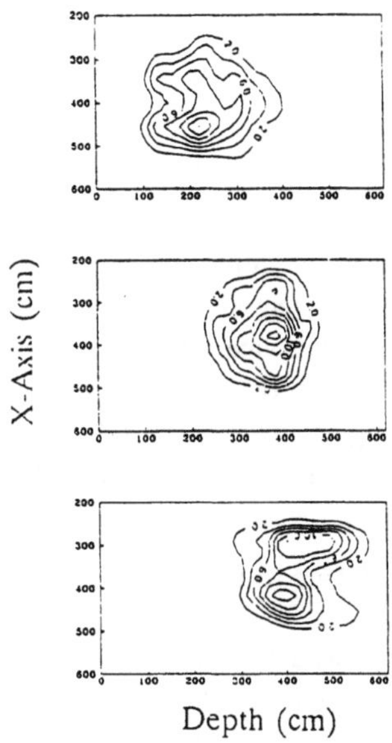

Figure 6b. A side-view of the solute plume in Figure 6a at the three times following injection (top 25, middle 48, and bottom 63 days)..

Clearly, the vertical change in the physical properties of the soil (i.e. texture and structure) need to be incorporated in models such as the CLT, which seem to be the most appropriate for describing transport near the soil surface. Our attempts to resolve the vertical correlation of solute velocity using static soil properties (e.g. Pb, θv) have been frustrated by deterministic trends overwhelming stochastic fluctuations. As a different approach, we suggest using the deterministic trend to formulate *a priori* the travel distance dependency of the CLT model parameters. As an illustrative example, assume that, under the quasi-steady state water flow conditions, the vertical volumetric water content profile reflects physical changes in the soil which are important in the movement of dissolved chemicals. At any depth z below a shallow calibration depth L, these changes in the soil properties may be related to the CLT parameters (μ and σ, the expectation and standard deviation of the natural log transformed solute travel time distribution) as:

$$\mu_z = \mu_L - \ln(R)$$
$$\sigma_z = \sigma_L \frac{zR}{L} \qquad (1)$$

where R is the ratio of the water content between 0 and L and 0 and z expressed mathematically as:

$$R = \frac{\int_0^L \theta(z)dz}{\int_0^z \theta(z)dz} \qquad (2)$$

In this model, the shape of the water content profile at steady state becomes a deciding factor in predicting solute dispersion. The effect of increasing water content with depth (reflecting an increase in fine soil separates) is to slow the solute velocity and to reduce the growth rate of the variance by narrowing the distribution of travel paths. For a vertically homogeneous soil, $\theta(z)$ is constant and Eq. (1) reduces to $\sigma_z = \sigma_L$ and $\mu_L + \ln (z/L)$. Unlike the linear growth of the dispersivity predicted by the CLT, the dispersivity (D_s) using Eq. (1) grows somewhat chaotically with travel distance z as shown in Figure 7 where,

$$D_s(z) = \frac{z}{2}\left[\exp\left(\sigma_z^2\right) - 1\right] \qquad (3)$$

The growth pattern displayed by Eq. (3) with its rapid rise, plateau, and eventual increase is similar to that observed but is not a convincing representation. To check the predictions of the BTC's, the value of μ_L in Eq. (1) may be determined from either the calibration depth BTC or by the piston flow expression for the mean travel time at L

$$\mu_L = \ln\left(\frac{L\theta}{q}\right) \qquad (4)$$

where q is the net applied water flux. Because the average travel time to 0.3 m was about twice as long as expected by piston flow (Butters et al., 1989) these two methods of assigning a value of μ_L yield very different results. Figure 8 shows the BTC's at 1.8, 3, and 4.5 m (solid line) and the original CLT model prediction (CLT 1). Included in the figure are the refined predictions using Eq. (1) and either Eq. (4) (CLT 2) or the BTC at L = 0.3 m (CLT 3) to determine μ_L. The CLT model prediction at 1.8 m is unchanged or worsened by the refined parameters. The model predictions at 3 m and 4.5 m are improved or worsened depending on the criteria used to select μ_L.

Although the results of the parameter adjustment on the CLT model predictions are mixed, it appears that knowledge of vertical changes in soil properties might be used to formulate a depth dependency in the mean and variance of the solute velocity without invoking geostatistical methods. The relationships suggested by Eq. (1) are consistent with observation at 3 and 4.5 m in this field but need to be tested in other soils. For wider application, formulations analogous to Eq. (1) using other soil properties like bulk density or soil texture might be tested. For example, replacing θ (z) in Eq. (2) with the percent (clay + silt) yields predictions which are very similar to those in Figure 8.

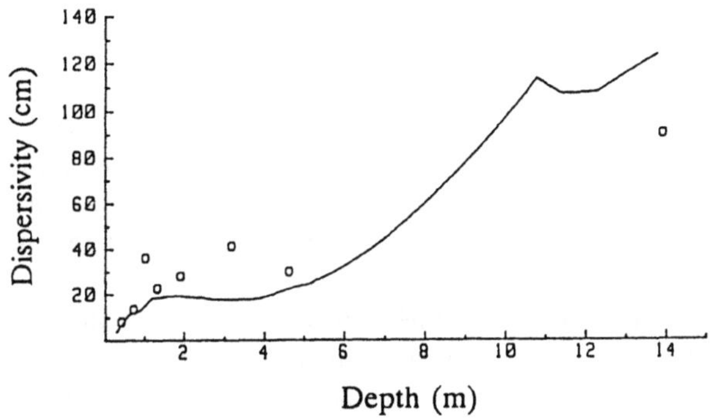

Figure 7. The observed macrodispersivity (O's) and that predicted using Eq. (1)–Eq. (3) (solid line).

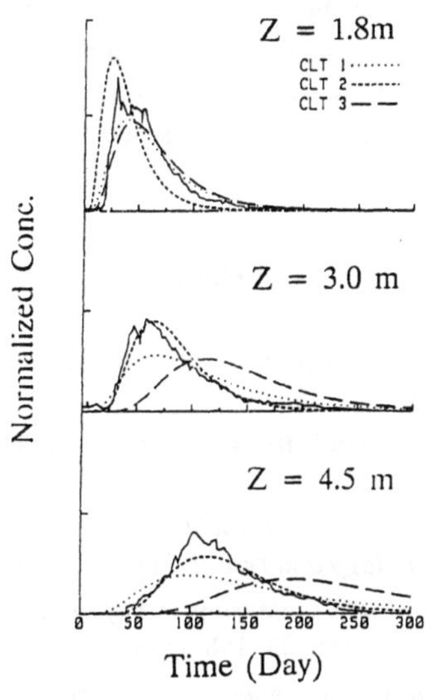

Figure 8. The normalized BTC's at 1.8 m, 3 m, and 4.5 m with the original CLT prediction (CLT 1) and the modified prediction using the water content profile and Eq. (1).

SUMMARY

The major findings of the sequence of studies may be summarized as follows:

1) In a vertically homogeneous, rapidly draining, sandy soil the stochastic convective lognormal transfer function model (CLT) provided an excellent simulation of observed solute movement under both steady state and transient flow conditions using a single, shallow calibration.
2) The dispersivity at a small scale (~ $3m^2$) within the field shows signs of approach to an asymptote by 14 m travel distance. Despite noticeable soil textural stratification, the differences between the local regions persist and the field scale dispersivity appears to increase without bound throughout the scale of observation.
3) The soil texture change near 3 m did not result in large scale lateral mixing but it did a) slow the advancing solute molecules allowing the trailing solute to catch up, and b) reduced the growth rate of the travel time variance by an unresolved mechanism.
4) A semi-empirical approach for adjusting the CLT model parameters in vertically heterogeneous soil was suggested. Using the water content profile, the CLT model predictions beyond a soil texture change may be improved or worsened depending on the method selected to calibrate the model.

References

Butters, G. L., 1987: *Field scale transport of bromide in unsaturated soil*, Ph.D. dissertation, University of California, Riverside.

Butters, G. L. and W. A. Jury, 1988: *Bromide transport through an unsaturated field soil*, report. Electr. Pow. Res. Inst., Palo Alto, Calif.

Butters, G. L., W. A. Jury, and F. F. Ernst, 1989: *Field scale transport of bromide in an unsaturated soil, 1. Experimental methodology and results*. Water Resour. Res., **25**, 1575-1581.

Butters, G. L. and W. A. Jury, 1989: *Field scale transport of bromide in an unsaturated soil, 2. Dispersion Modeling*. Water Resour. Res., **25**, 1583-1989.

Ellsworth, T., 1989: *Field scale spatial and temporal characterization of solute plume transport through unsaturated porous media*. Ph.D. dissertation, Univ. of Calif., Riverside.

Ellsworth, T. R., W. A. Jury, and F. F. Ernst, 1989: *A three-dimensional field study of solute transport through unsaturated layered, porousmedia; Methodology, mass recovery, mass transport*. Water Resour.Res. (In Review).

Jury, W. A., 1982: *Simulation of solute transport using a transfer functionmodel*. Water Resour. Res., **18**, 363-368.

Jury, W. A., L. H. Stolzy, and P. Shouse, 1982: *A field test of the transfer function model for predicting solute transport*. Water Resour. Res., **18**, 369-375.

Jury, W. A., J. S. Dyson, and G. L. Butters, 1989: *A transfer function model solute transport under transient water flow*. Soil Sci. Soc. Am. J. (In press).

Jury, W. A., G. L. Butters, L. D. Clendening, F. F. Ernst, H. Elabd, and T. R. Ellsworth, 1988: *Validation of solute transport models at the field scale*. In Wierenga and Bachelet (Eds.) Validation of flow and transport models for the unsaturated zone: Conference Proceedings, Ruidoso, New Mexico. Research Report 88-SS-04.

Reynolds, W. D., and D. E. Elrick, 1985: *In situ measurement of field-saturated hydraulic conductivity, sorptivity, and the alpha-parameter using the Guelph permeameter*. Soil Sci. **140**, 292-302.

Topp, G. C., and M. R. Binns, 1976: *Field measurement of hydraulic conductivity with a modified air-entry permeameter*. Can. J. Soil Sci. **56**, 139-147.

Greg L. Butters, Dept. of Agronomy, Colorado State University, Ft. Collins, CO 80523.
Tim Ellsworth, U.S. Soil Salinity Laboratory, Riverside, CA 92525.
William A. Jury, Dept. of Soil and Env. Sci., University of California, Riverside, CA 92521.

Field-Scale Water and
Solute Flux in Soils
Monte Verità

ASSESSMENT OF FIELD-SCALE LEACHING PATTERNS FOR MANAGEMENT OF NITROGEN FERTILIZER APPLICATION

D. J. Mulla and J. G. Annandale

Large-scale patterns in soil texture, saturated hydraulic conductivity, irrigation depth, and concentration of a bromide tracer were measured on a 57 ha commercial corn farm in Washington State, U.S.A. Texture of surface soil at the study site was predominantly a sandy loam. Saturated hydraulic conductivity was log-normally distributed with a mean of 13.8 cm/day. During the first month following corn germination, an average depth of 50 mm irrigation water was applied to the field. Bromide concentrations in the surface 0-15, 15-30, and 30-45 cm averaged 19.0, 4.2, and 2.9 ppm, respectively, after the third irrigation. These concentrations represent recovery of approximately 58% of the initial tracer application, the remainder being lost by leaching below the rooting zone. Geostatistics, joint frequency distributions, and a geographic information system were used to map and identify regions of the field having different rates of bromide leaching. Silt content of the surface soil and applied depth of irrigation water were used as criteria of the potential for leaching of bromide, with bromide leaching increasing as silt content decreased and irrigation depth increased. If uniformity of irrigation can be improved, the methods developed in this study offer good potential for assessing and managing broad patterns in leaching of nitrogen fertilizer.

1. Introduction

There is an increasing need in farming for fertility management systems which conserve natural resources, minimize the degradation of environmental quality, and reduce production costs. One of the most wasteful fertilizer practices is application of uniform rates and blends to large, spatially heterogeneous areas. Better methods are needed for identifying and dividing fields into small, homogeneous portions. Once the leaching regime and fertility status of each region is known, a variable rate or blend of nitrogen fertilizer can be applied to optimize soil fertility while minimizing the potential for contamination of ground water.

2. Objectives

The objectives of this study were i) to use geostatistics, joint frequency distributions, and a geographic information system to identify areas of a large commercial corn field having differing susceptibilities to solute leaching, and ii) to develop leaching management zones in which it would be possible to apply variable rates or blends of nitrogen fertilizer to small, homogeneous portions of the field to minimize the potential for leaching losses of fertilizer.

3. Methods

A 57 hectare irrigation circle located near George, Washington, U.S.A. was intensively sampled on a regular grid spacing of 30 m. At each site, three soil cores within a 5 m radius were collected to a depth of 45 cm and composited. All 603 samples were analyzed for percent sand, silt, and clay using the hydrometer method.

Bromide tracer was sprayed on 61 plots at a rate of 20 g/m^2. Each plot was 13 m^2 in area, and the separation between plots was 61 m. Plots were laid out along 5 transects, four of which were separated by 122 m. The fifth transect was perpendicular to the other four transects, and bisected them at their midpoint.

At each plot, measurements of saturated hydraulic conductivity were made using a Guelph permeameter (Reynolds and Elrick, 1985). Depth of sprinkler irrigation at each plot was measured using catch cans. After each irrigation, three soil cores were collected and composited at depths of 0-15 cm, 15-30 cm, and 30-45 cm. Bromide was extracted by 30 min. of shaking 25 g soil with 50 ml of an aqueous sodium nitrate solution and subsequently allowing soil particles to settle out of solution (Abdalla and Lear, 1975). The supernatant was analyzed for bromide concentration using a specific ion electrode.

For the purposes of computing mass recovery, solution bromide concentrations were doubled to obtain concentration per unit mass of dry soil. These concentrations were summed over the three soil depths and multiplied by average soil bulk density (1.65 g/cm^3) and by the depth of sampling (15 cm) to obtain total mass of bromide recovered in the depths sampled. Mass recovery expressed as a fraction is simply the latter result divided by the initial mass of bromide applied in a square meter (20 g).

Corn was sown in the field on May 26, 1989. All 61 plots were sprayed with bromide on June 1. Irrigation was applied at the discretion of the farmer, without consideration of potential evapotranspiration rates. The first sprinkler irrigation occurred from June 6-7. The first samples for bromide were taken on June 8. The second and third sprinkler irrigations were on June 13-14 and June 22-23. The second soil samples were taken on June 26.

4. Results and discussion

Soils at the study site had an average textural classification of sandy loam (Table 1), but portions of the field had a loamy sand texture. Due to the high sand content of soils, the mean saturated hydraulic conductivity was large (14 cm/day). The total depth of irrigation (Table 1) applied prior to the second sampling for bromide averaged 50 mm. Due to wind and non-uniform sprinkler efficiencies, depth of irrigation exhibited a moderate amount of spatial variability with a coefficient of variation of about 30%. Prior to the second bromide sampling, corn plants had small leaf area indices (<0.34), indicating small canopies and rooting systems.

Table 1: Statistical data for measured soil properties, irrigation depth, and corn crop development.

Measured variable	mean	S.D.	C.V.
sand (%)	68.8	6.3	9.1
silt (%)	21.0	4.4	20.9
clay (%)	10.0	2.9	29.0
vol. water content at "field capacity" (%)	25.9	3.8	14.7
irrigation depth: first cycle (mm)	23.0	7.0	30.4
second plus third cycle (mm)	27.0	9.0	33.3
saturated hydraulic conductivity (cm/day)	13.8	16.9	122.5
leaf area index (June 5)	0.006	---	---
leaf area index (June 21)	0.34	---	---

Concentrations of bromide (Table 2) in the soil were highest in the 0-15 cm increment at both sampling times. Bromide levels at the 15-30 cm depth were from 4-5 times lower than levels at the surface. Bromide was detected at the 30-45 cm depth in low concentrations. Variability in bromide concentrations was moderate to high in magnitude, and comparate to values observed by Butters et al. (1989). The average recovery of bromide in the 0-45 cm profiles sampled was 86% on the first sampling and 65% on the second sampling. It was assumed that plant uptake of bromide was negligible due to the small size and leaf area of plants. Thus, the remainder of bromide was most likely lost by leaching below the 45 cm depth. Analysis of bromide in plant tissue on the third bromide sampling (July 13) showed that nearly all bromide remaining in the 0-45 cm soil depth on the second sampling was taken up by plants. Leaf area index of plants on the third sampling averaged 3.6, indicating full development of corn canopy and rooting system.

Table 2: Statistical data for soil extract bromide concentrations measured in 61 plots at three depths on two dates.

Bromide	mean*	S.D.	C.V.
first sampling:			
0-15 cm depth (ppm)	24.7	6.9	27.9
15-30 cm depth (ppm)	6.0	4.3	71.7
30-45 cm depth (ppm)	4.2	4.0	95.2
second sampling:			
0-15 cm depth (ppm)	19.0	8.2	43.1
15-30 cm depth (ppm)	4.2	2.0	47.6
30-45 cm depth (ppm)	2.9	1.9	65.5

* multiply by 2 to express bromide concentrations on a dry soil mass basis.

Plots having low concentrations of bromide at the 0-15 cm depth represent areas with higher leaching losses than plots having high bromide levels. To identify which soil factors may control leaching of bromide, a correlation matrix (Table 3) between bromide in the surface increment and other measured properties was computed. The highest correlations were found between bromide and either silt content or depth of irrigation. These results indicate that plots having lower silt content had lower bromide concentrations, and plots which received lower rates of irrigation had higher bromide concentrations. The first irrigation cycle was only weakly correlated with bromide concentrations in the surface (r=0.11). At the time of the first irrigation, soil had been recently tilled and was relatively dry. Thus, the main effect of the first irrigation was to leach the bromide at the soil surface into the surface 0-15 cm depth, and little deep leaching losses resulted.

Table 3: Simple correlation coefficients (r) between concentrations of bromide at 0-15 cm on the second sampling date and other measured data.

Variable	Bromide (ppm)
sand (%)	-0.27*
silt (%)	0.31*
clay (%)	0.02
saturated hydraulic conductivity (cm/day)	-0.02
irrigation depth on the second plus third cycle (mm)	-0.30*

*significant at a 5% confidence level.

Kriged spatial patterns in percent silt and cumulative depth of the second plus third irrigations are shown in Figures 1 and 2, respectively. Both variables obeyed gaussian frequency distributions. Based upon the frequency distributions, two cutoff values defining low, medium, and high levels of percent silt or depth of the second plus third irrigations were established. Together, these three pairs of ranges in data can be combined into nine possible combinations representing a joint frequency distribution between silt content and irrigation depth (Table 4).

Table 4: Area covered by nine possible groupings of the joint frequency distributions for percent silt and depth of the second plus third irrigations (mm).

Leaching category		Area (ha)
#1: SILT<19	& IRRIG<20	1
#2: SILT<19	& 20≤IRRIG<30	8
#3: 19≤SILT<25	& IRRIG<20	3
#4: 19≤SILT<25	& 20≤IRRIG<30	21
#5: SILT<19	& IRRIG≥30	6
#6: SILT≥25	& IRRIG<20	1
#7: 19≤SILT<25	& IRRIG≥30	8
#8: SILT≥25	& 20≤IRRIG<30	8
#9: SILT≥25	& IRRIG≥30	0

Each of the resulting nine categories is termed a leaching category in that the combination of percent silt and depth of irrigation affects the potential leaching losses of bromide.

For example, leaching category #4, which covers 21 ha, represents regions having both moderate silt content and moderate irrigation. This category would be expected to have moderate leaching losses of bromide. Category #5, which covers 6 ha, would be expected to have high leaching losses, since silt content is low and irrigation is high.

A geographic information system (Berry, 1986) was used to combine regions of the field having the potential for low, medium, or high leaching losses of bromide (Table 5). These combinations are termed leaching management zones. Low leaching zones are characterized by high silt content and low irrigation depth. High leaching zones are characterized by low silt content and high irrigation. The low, medium, and high leaching zones cover areas of 13, 21, and 23 ha, respectively. Bromide concentration in the 0-15 cm depth is lowest in the high leaching zone and is highest in the low leaching zone. This is qualitatively the type of result that would be expected, and confirms the validity of using joint frequency distributions to identify and divide a field into small homogeneous portions having different leaching regimes.

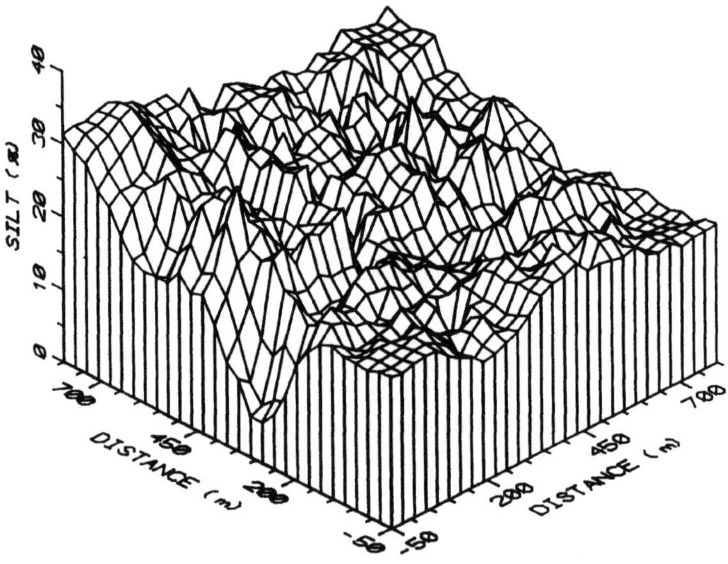

Figure 1: Spatial pattern in percent silt content

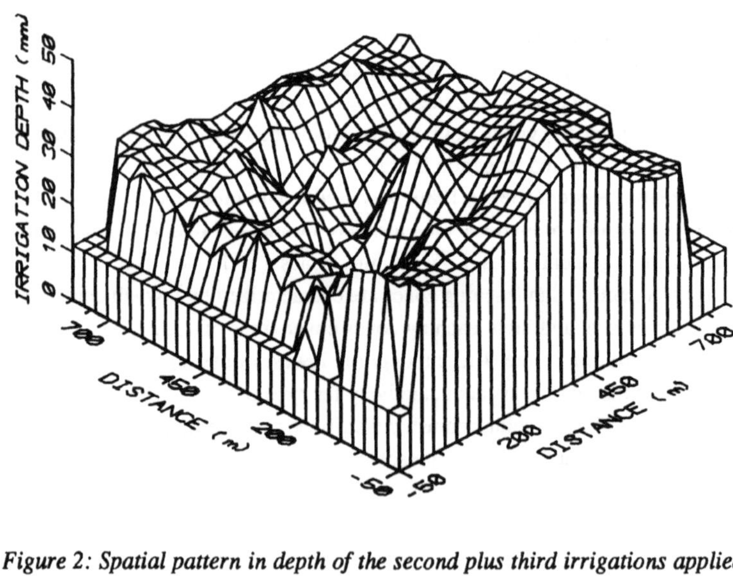

Figure 2: Spatial pattern in depth of the second plus third irrigations applied prior to the second sampling for bromide

Table 5: Leaching management zone groupings, and the leaching categories in each zone. Mean values for percent silt, depth of the second plus third irrigations, bromide concentration, and area in each leaching management zone are also reported.

Leaching management zone & leaching categories in zone	Silt (%)	Irrig. depth (mm)	[Br⁻] (ppm)	Area (ha)
Low leaching zone (#1, #3, #6, #8)	25.3	22.5	24.9	13
Med. leaching zone (#4)	21.8	25.0	19.6	21
High leaching zone (#2, #5, #7)	18.6	32.3	18.4	23

A graphical representation of the three leaching zones from the analysis in Table 5 is shown in Figure 3. A leaching index of 5, 15, or 25 is used to represent low, medium, and high leaching management zones, respectively. Most of the field in the foreground of Fig. 3 has a leaching index of 25, and represents regions in the high leaching management zone. A map of surface concentrations of bromide is shown in Fig. 4. Areas with lower concentrations of bromide generally match areas in Fig. 3 in the high leaching management zone. Areas with higher bromide levels generally correspond to the areas in the low leaching management zone. Thus, a satisfactory inverse correspondence exists between the leaching management zone map and the map of surface bromide levels.

A leaching management zone map is potentially useful for customized nitrogen fertilizer applications that are designed to minimize contamination of groundwater. Mulla (1989) has demonstrated how applications of phosphorus fertilizer can be varied using similar mangement maps and computerized spreading equipment with field locator capabilities. For the case of the relatively mobile nutrient nitrate, it is desirable to prevent rapid eaching below the rooting zone. To achieve this, portions of the field residing in the high leaching management zone could be managed with lower applications rates of nitrogen fertilizer than portions in the low or medium leaching zones. Alternatively, the high leaching zone could receive a blend of nitrogen fertilizer that is relatively immobile; i.e. urea with a nitrification inhibitor. Further research is needed to evaluate the feasibility of such an approach.

To effectively reduce nitrate nitrogen leaching losses on irrigated sandy soil, it is important that nitrogen fertilizer applications closely match crop yield requirements (Hergert, 1986). In additions, irrigation scheduling is necessary to prevent excess leaching. In the present study, it was observed that spatial variation in depth of applied irrigation had a significant influence on bromide leaching losses. We suggest that attention should be given to increasing irrigation uniformity so that management decisions relating to leaching can be based primarily on the basis of spatial patterns in soil texture.

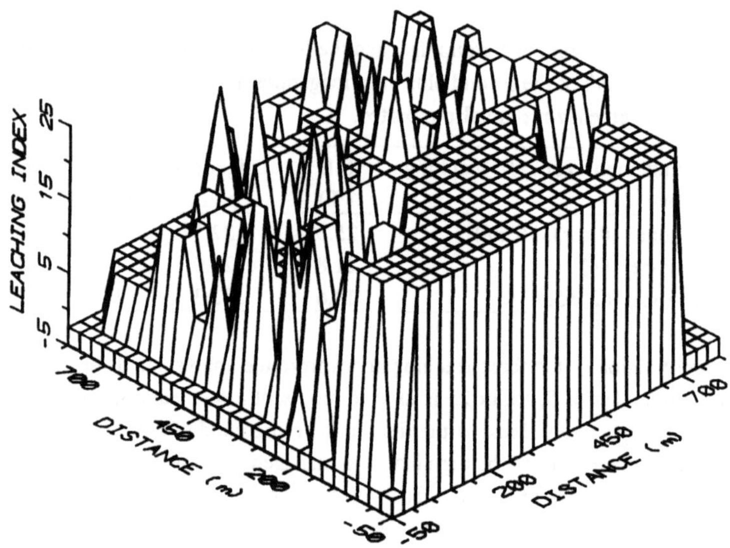

Figure 3: Location of low (index=5), medium (index=15), and high (index=25) leaching management zones

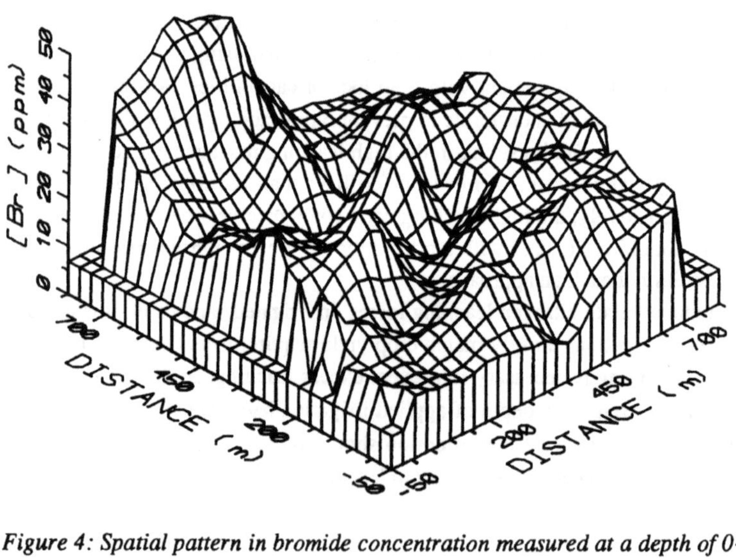

Figure 4: Spatial pattern in bromide concentration measured at a depth of 0-15 cm on the second sampling date.

5. Conclusions

Large-scale patterns in soil texture, irrigation depth, and concentration of a bromide tracer were measured on a 57 ha commercial corn farm in Washington state, U.S.A. Texture of surface soil at the study site was predominantly a sandy loam. During the first month following germination, an average depth of 50 mm of irrigation water was applied to the field. Bromide concentrations in the surface 0-15, 15-30, and 30-45 cm averaged 19.0, 4.2, and 2.9 ppm, respectively, for an average recovery of 58%. The remainder of bromide was assumed lost by leaching below the rooting zone, because leaf area index was too small for significant uptake of bromide by plants. Surface bromide concentrations were lowest in regions of low silt and high irrigation, and a significant statistical correlation existed between bromide and both soil texture and irrigation depth. A joint frequency distribution of silt content and irrigation depth was used to divide the field into leaching categories, and these categories were subsequently grouped into leaching management zones using a geographic information system. Bromide concentration in surface soil was highest in the low leaching zone (covering 13 ha) and lowest in the high leaching zone (covering 23 ha). To reduce the potential for contamination of ground water, it is suggested that the high leaching zone be managed with lower application rates of nitrogen fertilizer or with relatively immobile formulations of nitrogen.

References

Abdalla, N.A. and B. Lear, 1975: *Determination of inorganic bromide in soils and plant tissues with a bromide selective-ion electrode.* Commun. Soil Sci. Plant Anal., **6** (5), 489-494.

Berry, J.K., 1986: *GIS: Learning computer-assisted map analysis.* J. Forestry, **84** (10), 39-43.

Butters, G.L., W.A. Jury, and F.F. Ernst, 1989: *Field scale transport of bromide in an unsaturated soil. 1. Experimental methodology and results.* Water Resour. Res., **25** (7), 1575-1581.

Hergert, G.W., 1986: *Nitrate leaching through sandy soil as affected by sprinkler irrigation management.* J. Environ. Qual., **15** (3), 272-278.

Mulla, D.J., 1989: *Soil spatial variability and methods of analysis.* In: C. Renard, R.J. Van Den Beldt, and J.F. Parr (eds.), Soil, Crop, and Water Management Systems for Rainfed Agriculture in the Sudano-Sahelian Zone, ICRISAT, Patancheru, India.

Reynolds, W.D., and D.E. Elrick, 1985: *A laboratory and numerical assessment of the Guelph permeameter method.* Soil Sci., **144** (4), 282-299.

D. J. Mulla and J. G. Annandale, Department of Agronomy and Soils, Washington State University, Pullman, Washington 99164 6420, U.S.A.

Field-Scale Water and
Solute Flux in Soils
Monte Verità

THE EFFECT OF FIELD SOIL VARIABILITY IN WATER FLOW AND INDIGENOUS SOLUTE CONCENTRATIONS ON TRANSFER FUNCTION MODELLING OF SOLUTE LEACHING

R.E. White and L.K. Heng

The transport of soil or fertilizer-derived nutrient ions, chloride, nitrate and sulphate, through defined soil volumes under the influence of surface-applied water is treated as a stochastic process. The probability density function of travel times of externally applied chloride, or of indigenous solutes, was used to define the volume of soil effective in solute transport during individual leaching events, for intact cores in the laboratory or mole-and-tile drained soils in the field. The fractional transport volume θ_{st} (m^3 m^{-3}) of a soil was little affected by variable rates of water input or the initial moisture status. Thus, a soil's water flow characteristics could be reasonably well defined from the ensemble behaviour of intact cores, or the integrating effect of an underdrainage system. The major uncertainty in predicting quantities of solute leached, however, lay in estimating the initial concentration of solute in the soil's transport volume. When the value of this initial condition is known accurately, good agreement between predicted and measured quantities of solute leached may be obtained.

1. Introduction

The movement of surface-applied or indigenous solutes in solution through soil has been studied for at least 100 years (Lawes et al., 1882; Nielsen et al., 1986). Interest has quickened in recent years, however, because of widespread public concern that undesirable chemicals and waste materials are leaching from soil and polluting underground and surface waters.

Attempts to model this movement quantitatively have generally involved combining equations for fluid flow through porous media with equations for solute travel by mass flow and diffusion. This deterministic approach works well for conservative solutes that are transported by steady water flow through homogeneous soil. In practice, however, soil scientists are confronted by an essentially chaotic system: rainfall inputs that are episodic and variable in intensity, flow conditions ranging from transient to steady-state, complex and heterogeneous conducting pathways for water, and solutes that are reactive, biologically labile and can vary in concentration both spatially and temporally within the soil volume.

In an effort to provide utilitarian models which can cope with field soil heterogeneity and the variability of natural processes, some have adopted a stochastic treatment of solute transport in soil. One example is the transfer function approach of Jury (1982), Jury et al. (1982, 1986) and White (1987, 1989), which depends on a knowledge of the travel time distribution of a solute passing from a defined input surface to an output surface in the soil. The resultant transfer function model (TFM) has much in common with the mass response function (MRF) developed for drainage basin analysis of contaminant responses to rainfall pulses (Rinaldo et al., 1989). Despite the reservations of "determinists" that stochastic modelling virtually abandons any dependence on the physics of water flow (Towner, 1989), the approach has much potential for simulation of field-scale solute transport. However, as Nielsen et al. (1986) point out, extensive experimental verification of the models is necessary before they can be employed with confidence.

2. Model development and evaluation

This paper briefly reviews recent experiments conducted on drained field soils, and large intact cores taken from those soils, which had the aim of developing and evaluating a transfer function model for predicting the leaching of nitrate and sulphate. Sites under pasture were chosen on an Evesham clay loam soil (Aquic Eutrochrept) at Wytham, near Oxford, England and on a Tokomaru silt loam soil (Typic Fragiaqualf) at Palmerston North, New Zealand. Both soils had well developed granular to subangular blocky structure in the A horizon, grading into an angular blocky to prismatic structure in the B horizon. Full details of the experimental procedures are given in White et al. (1986), Haigh and White (1986) and White (1987, 1989).

The general form of the transfer function equation used was:

$$Q_{ex}(t) = \int_{o}^{t} g(t-t'|t')_L \, Q_{ent}(t')dt' \qquad (1)$$

where $Q_{ent}(t')$ and $Q_{ex}(t)$ are, respectively, the mass rate of solute input to the soil at time t' and the mass rate of solute output at the depth of interest (L) at time t. $g(t-t'|t')_L$ is the probability density function (pdf) of solute travel times (t-t') between the surface and depth L, conditional upon solute input at time t'. Within the experimental soil volume, there is a fluid volume which appears to be effective in transporting solute during the time scale of observation. This transport volume V_{st} is likely to be highly irregular in shape and variable in size because of the complexity of water movement, especially under transient unsaturated conditions, in a heterogeneous porous body such as a structured soil. Nevertheless, one of the objectives of this research was to determine to what extent the operationally defined parameter V_{st} for a particular soil volume, or more generally the fractional transport volume θ_{st} (where $\theta_{st} = V_{st}$/soil volume), was a characteristic of a given soil type, under defined management conditions, that could be used for the predictive modelling of solute leaching using relatively easily gathered input parameters (cf. Rinaldo et al., 1989).

The experiments involved step inputs of chloride to the surface of soil cores and pulse inputs of chloride and sulphate to field soils under intermittent rainfall. Where possible, the pdf of travel times for the externally applied chloride was used to characterize the behaviour of the soil system, and as a basis for modelling the movement of labile NO_3^- and SO_4^{2-} already resident in the soil.

3. Experiments with intact cores

Chloride transport

Thirteen intact cores of Evesham soil, each 23 cm in diameter and 28 ± 0.2 cm long, were taken on a grid pattern from 1 ha of grass pasture at Wytham. Each core was leached with 10 mM $CaCl_2$ at a constant flux density (q_o) of 7.2 ± 0.3 mm h^{-1} until approximately 1 pore volume (c. 5 l) of effluent had been collected. A comparison of volume rates of inflow and outflow showed that flow through the cores was unsteady for one-third to one-half of the whole leaching period (White, 1989). The base of each core was open to the atmosphere, as Dyson and White (1987) had shown this to be a reasonable lower boundary condition for the simulation of field leaching conditions in this soil. Individual effluent samples (ranging from 50 to 500 ml) were analysed for Cl^- and NO_3^-.

From eqn (1), when $Q_{ent}(t')$ for Cl is constant, we have:

$$[P_L(t)]_{Cl} = \frac{Q_{ex}(t)}{Q_{ent}(t')} = \int_o^t g(t-t'|t')_L dt' \qquad (2)$$

where $[P_L(t)]_{Cl}$ is the probability that a Cl ion, having entered the soil's transport volume at time t', will have exited at depth L in time t. If each breakthrough curve for chloride were treated as a single sample realization of the leaching process, the 13 observations could be pooled to produce an "ensemble" probability plot of $P_L(t)$ against time (Fig. 1). On the evidence that the pdf of solute travel times in field soils (Jury et al., 1982; White, 1987) and intact cores (White 1985a; Dyson and White, 1987) conformed to a lognormal function, the integral on the right-hand-side of eqn (2) could be evaluated and the equation written as:

$$[P_L(t)]_{Cl} = 0.5\left\{1 + \text{erf}\left[\frac{\ln t - \mu_L}{\sqrt{2}\sigma_L}\right]\right\} \qquad (3)$$

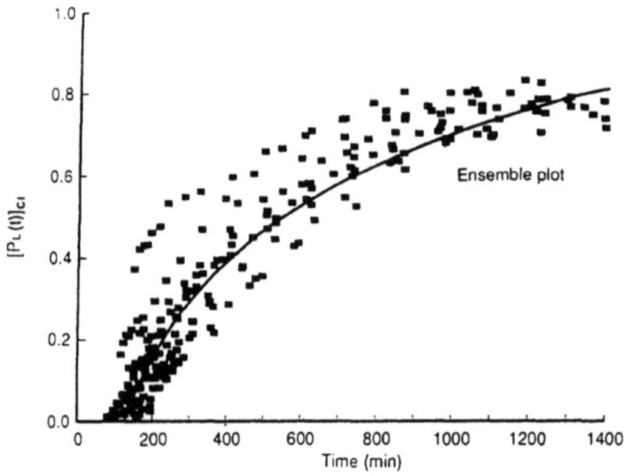

Figure 1: A plot of probability $[P_L(t)]_{Cl}$ against observation time for 13 Evesham soil cores (after White, 1989).

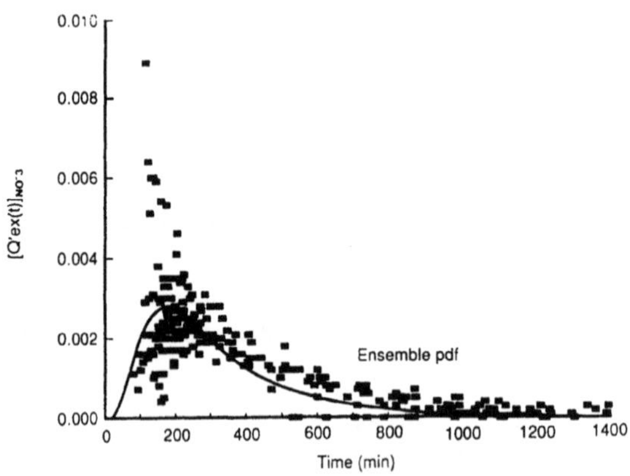

Figure 2: A plot of the normalized mass rate of nitrate efflux $[Q'_{ex}(t)]_N$ against observation time for 13 Evesham soil cores (after White, 1989).

where L was the mean core length (28 ± 0.2 cm) and μ_L and σ^2_L, were the first and secon‿ moments of the lognormal distribution. Equation (3) was fitted to the data of Fig. 1 by least squares optimization to give best-fit values of μ_L and σ_L, which are given in Table I. The centre of location of the travel time distribution was chosen as the median travel time t_m because it could be unequivocally calculated as the time corresponding to $[P_L(t)]_{Cl} = 0.5$ in eqn (3). The value of t_m for Cl^- was then given by $\exp(\mu_L)$ and the "ensemble average" fractional transport volume θ_{st} for Cl^- was calculated from the equation (White et al., 1986):

$$\theta_{st} = q_o t_m / L \tag{4}$$

Nitrate transport

Additionally, the Cl effluent data can be used to model the leaching of indigenous soil NO_3^-, following the premise that the probability P(t) that a Cl^- ion, having entered the transport volume V_{st} at a time t', will remain in V_{st} at a later time t, is related to the probability that the ion has exited from V_{st} through the lower surface at depth L by the equation (White, 1989):

$$P(t) = 1 - [P_L(t)]_{Cl} \tag{5}$$

Table I: Best-fit values of μ_L and σ_L, assuming a lognormal distribution of solute travel times, and the estimated value of θ_{st} for surface-applied chloride in an assembly of Evesham soil cores.

Parameters of the lognormal pdf		Fractional transport volume
μ_L	σ_L	θ_{st} ($m^3\ m^{-3}$)
6.3069 ± 0.0209	1.0462 ± 0.0254	0.235 (0.219-0.252)*

* Range for 1 S.E. above and below the mean using error estimates for all the variables in eqn(4)

Initially, the probability of a surface-applied Cl-ion being retained within the soil's transport volume is high, but decreases as the observation time lengthens and the value of $[P_L(t)]_{Cl}$ increases. The behaviour of an initially resident NO_3^- ion should be analogous to that of a Cl- ion, once it has entered V_{st}, so that the left-hand-side of eqn(5) can be replaced by the concentration ratio $C_r(t)/C_i$ where $C_r(t)$ and C_i are the volume-averaged concentrations of NO_3^- in the transport volume at time t and time zero, respectively. If it is assumed that NO_3^- losses from the soil occur only via the transport volume, then $C_r(t)$ may be replaced by $C_f(t)$, the flux-averaged concentration of NO_3^- in the effluent from each core. Equation (5) can then be rewritten as:

$$\left[\frac{C_f(t)}{C_i}\right]_N = 1 - \left[\frac{Q_{ex}(t)}{Q_{ent}(t')}\right]_{Cl} \tag{6}$$

Given the Cl breakthrough data of Fig. 1, eqn(6) predicts that $C_f(t)$ for NO_3^- should show a monotonic decrease with time. Such behaviour was not observed in this, or previous leaching experiments (White, 1985a). Further, the mass rate of NO_3^- efflux $[Q_{ex}(t)]_N$, derived from eqn(6) by multiplying through by $C_i\, i_{ex}(t)$ to give:

$$[Q_{ex}(t)]_N = C_f(t) i_{ex}(t) = C_i i_{ex}(t) \left\{ 1 - \left[\frac{Q_{ex}(t)}{Q_{ent}(t')} \right]_{Cl} \right\} \tag{7}$$

where $i_{ex}(t)$ is the volume rate of outflow at time t, showed a pronounced maximum some time after the start of leaching (see Fig. 2, White, 1989). When the mass rate of NO_3^- efflux for each core was normalized for the total amount of NO_3^- leached from each core, and the values, $[Q'_{ex}(t)]_N$, plotted against time (Fig. 2), this unexpected trend in the rate of NO_3^- leaching was confirmed.

The form of the $[Q'_{ex}(t)]_N$ vs time plots suggested that an alternative approach to modelling the leaching of soil NO_3^- might be to assume a single pulse of NO_3^- was present in, or moved into, the transport volume at the start of leaching, and was subsequently transported by the flux of surface-applied water through the soil into the effluent. Applying this analysis to eqn(1), the input flux of NO_3^- to the transport volume becomes the product of a Dirac "delta-function" $\delta(t')$ and a normalization constant k (equal to the total mass of NO_3^- leached during one leaching experiment), and the integral in eqn(1) collapses to give the equation:

$$[Q_{ex}(t)]_N = k\,[g(t)]_N \tag{8}$$

where $[g(t)]_N$ is the pdf of travel times of indigenous soil NO_3^-. $[g(t)]_N$ is therefore the same as $[Q'_{ex}(t)]_N$ and the plots for 13 soil cores in Fig. 2 provide an "ensemble" pdf of NO_3^- travel times.

Assuming a lognormal distribution of travel times (as for Cl^-), the data in Fig. 2 were fitted by least squares optimization using the equation:

$$[Q'_{ex}(t)]_N = [\sqrt{2\pi}\,\sigma t]^{-1} \exp\left[-(\ln t - \mu)^2/2\sigma^2 \right] \tag{9}$$

Although narrow in the time sense, the pulse of indigenous NO_3^- can be treated, in the absence of contrary information, as being distributed over the whole surface of the transport volume, excluding the exit surface (see Jury et al., 1986). The mean travel distance of NO_3^- ions to the exit surface was therefore taken as L/2 (= 14 cm). Values of $\mu_{L/2}$ and $\sigma_{L/2}$ obtained from eqn(9), and of θ_{st} for NO_3^- obtained from eqn(4), are given in Table II.

The close agreement between θ_{st} for Cl^- (Table I) and θ_{st} for NO_3^- (Table II) suggested that the space and time-averaged transport volumes for the two solutes were comparable in these leaching experiments. Further evidence of the similar behaviour of NO_3^- and Cl^- was provided by an analysis of the first moments of their lognormal pdfs. When the travel times of a solute moving between two depths, L/2 and L, in the soil are perfectly correlated, we may write (Jury, 1982):

$$\mu_L = \mu_{L/2} + \ln 2 \tag{10}$$

Substituting μ_L for Cl^- in eqn(10) and solving for $\mu_{L/2}$ gives a value of 5.614, which is very close to the best estimate of $\mu_{L/2}$ for NO_3^- shown in Table II. However, $\sigma_{L/2}$ for NO_3^- was considerably less than σ_L for Cl^- which may be a consequence of treating the influx of NO_3^- to the transport volume as a narrow pulse at the start of leaching.

Table II: Best-fit values of $\mu_{L/2}$ and $\sigma_{L/2}$, assuming a lognormal distribution of solute travel times, and the estimated value of θ_{st} for resident NO_3^- in an assembly of Evesham soil cores.

Parameters of the lognormal pdf		Fractional transport volume
$\mu_{L/2}$	$\sigma_{L/2}$	θ_{st} (m^3 m^{-3})
5.593 ± 0.0327	0.6817 ± 0.0282	0.230 (0.212 - 0.248)*

* Range for 1 S.E. above and below the mean using error estimates for all the variables in eqn(4)

The θ_{st} value of 0.230 to 0.235 was less than half the mean water-filled pore volume for the 13 cores of 0.527 ± 0.007, a result consistent with that obtained previously by Dyson and White (1987) for 18 cores of mean length 16.4 cm, leached under steady-state, non-ponding conditions. This suggested that considerable "by-pass" flow occurred in the Evesham soil.

4. Experiments with mole-and-tile drained field soils

Concentrations of NO_3^- and Cl^- in the drain flow from a mole-and-tile drained field of 1.1 ha at Wytham near Oxford were monitored using automatic water sampling equipment for individual rain events during several seasons (White, 1987). Assuming the input of solute mass (indigenous NO_3^- or Cl^-) to the soil's transport volume occurred during a short period at the start of each event, the efflux data were analysed following eqns (1), (8) and (9), and a lognormal function fitted to the plots of $Q'_{ex}(t)$ against time for both NO_3^- and Cl^-. Median travel times t_m, calculated from the best-fit values of μ, were composites of a travel time through the soil to the mole drains, and a travel time through the mole-and-tile drain system to the collection point. However, the median travel time for the latter part of the system (t_d) was < $^1/_{20}$th that of the overall travel time, so the median travel time through the soil to drain depth could be calculated as $t_m - t_d$. Recalling the previous discussion in respect of the distribution of indigenous Cl^- and NO_3^- in the soil at the start of each event, the distance of travel for these ions was taken as L/2, where L was the average drain depth of 32.5 cm. Values of θ_{st} for Cl^- and NO_3^- were calculated by substituting values of $t_m - t_d$ in eqn(4).

The results in Table III show that θ_{st} was comparable for Cl^- and NO_3^- for each event, and did not change appreciably from event to event in spite of variations in the rainfall intensity and duration. The events were monitored over several seasons, but θ_{st} did not seem to be particularly sensitive to variations in initial soil moisture content within the range 0.45 to 0.55 m^3 m^{-3} that

Table III: θ_{st} values based on median travel times for soil chloride and nitrate in the mole-and-tile drained Evesham soil for 11 rainfall events of varying intensity and duration.

Date of event	Rainfall duration (h)	Mean rainfall intensity, q_o (mm h^{-1})	θ_{st} (m^3 m^{-3}) Cl^-	NO_3^-
21/11/82	13	0.88	*	0.080
31/1/83	7	1.74	0.098	0.092
21/5/83	3	3.93	0.200	0.170
1/6/83	6	2.43	0.074	0.062
16/1/84	8	1.10	0.068	0.070
6/2/84	3	1.27	0.024	0.030
23/3/84	4	1.75	0.070	0.072
29/11/85	8	1.30	0.140	0.120
2/12/85	5	1.08	0.064	0.064
4/12/85	22	0.68	0.088	0.088
6/12/85	6	1.17	0.086	0.084
Mean values of θ_{st} (± S.E.) for all events			0.092 ± 0.015	0.084 ± 0.011

* Cl transport not measured

was experienced. This suggested that a space and time-averaged transport volume could be used to characterize a particular soil type, which would allow predictions of field-scale solute leaching to be made. For example, instead of requiring the transfer function equation(1) to be integrated over time for individual events to obtain the quantity of solute leached (ΔS), when the space and time-averaged transport volume V_{st} is known, the approximate equation developed by White (1987, 1989)

$$\Delta S = C_i V_{st}\left(1 - \exp\left(- V_{ex}/V_{st}\right)\right) \tag{11}$$

can be used. Here C_i is the mean concentration of solute in the transport volume initially and V_{ex} is the cumulative volume of drainage for an event. Measured and predicted amounts (derived from eqn(11)) of N leached for the mole-and-tile drained Evesham soil are compared in Fig. 3, which shows good agreement for 10 of the 11 events.

The utility of this approach is demonstrated further by results obtained for Cl^- and SO_4^{2-} leaching in mole-and-tile drained fields of the Tokomaru silt loam at Palmerston North. Two hydrologically similar paddocks, each 0.125 ha in area, were treated with KCl (200 kg/ha) and S (50 kg/ha), applied as elemental S to one paddock and SO_4-S to the other. Several drainage events were monitored during the winter of 1988, and Cl^- and SO_4^{2-} concentrations were measured in the drain flow. The plots of $Q'_{ex}(t)$ vs time for Cl^- and SO_4^{2-} were modelled using lognormal

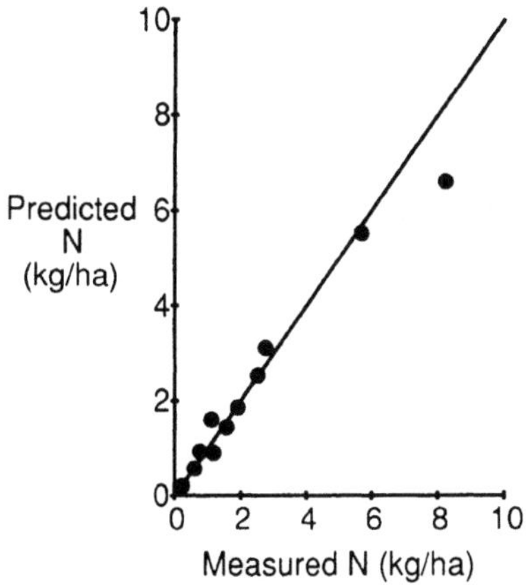

Figure 3: A comparison of predicted (using eqn. 11) and measured N leached for the drained Evesham soil (solid line represents a 1:1 ratio).

functions to obtain estimates of μ_L and σ_L for each event, where the drain depth L was 45 cm. Median travel times and θ_{st} values were calculated as before. The results, summarized in Table IV, show consistently good agreement between θ_{st} values calculated from Cl^- or SO_4^{2-} transport, irrespective of rainfall event or paddock, except for the event of 6 October 1988. When the quantities of Cl^- of SO_4^{2-} leached were calculated from eqn(11) using V_{st} for chloride, and compared with measured quantities of Cl^- and SO_4^{2-} leached, reasonable agreement was obtained over the whole range of values (Fig. 4).

It is also noteworthy that when θ_{st} values for the drained Tokomaru soil were averaged over the winter season, the means of 0.056 ± 0.004 and 0.075 ± 0.015 $m^3\ m^{-3}$ for Cl^- and SO_4^{2-}, respectively, were similar to the space and time-averaged θ_{st} values of 0.084 for NO_3^- and 0.092 m^3m^{-3} for Cl^- calculated for the Evesham field soil (Table III). These θ_{st} values were low compared to soil volumetric water contents which were in the range 0.45 to 0.55 $m^3\ m^{-3}$ for the Evesham soil, and 0.4 to 0.5 $m^3\ m^{-3}$ for the Tokomaru soil, suggesting a predominance of preferential flow through macropores and fissures in the soil over the mole channels. Similarly, White et al. (1986) calculated θ_{st} values of 0.067 to 0.106 m^3m^{-3} for bromide movement from the soil surface to individual mole drains in the Evesham soil. It is tempting to speculate, therefore, that the transport characteristics of mole-drained soils are determined primarily by the fissuring induced by the mole plough, and the subsequent structural development in the moled soil, especially in the regions over the moles (which are normally about 2 m apart).

Table IV: θ_{st} values based on median travel times following surface inputs of chloride and sulphate in the mole-and-tile drained Tokomaru soil for 6 rainfall events of varying intensity and duration.

Date of event	Paddock*	Rainfall duration (h)	Mean rainfall intensity, q_o (mm h^{-1})	θ_{st} Cl^- (m^3 m^{-3})	θ_{st} SO_4^{2-} (m^3 m^{-3})
18/6/88	A	12	2.03	0.045	0.046
	B			0.044	0.043
8/8/88	A	6	2.95	0.050	0.047
	B			0.060	0.057
20/8/88	A	8	1.55	0.060	0.056
	B			0.055	0.055
24/8/88	A	14	1.23	0.045	0.047
	B			0.055	0.053
21/9/88	A	4	3.75	0.069	0.046
	B			0.072	0.060
6/10/88	A	2	7.75	0.228	0.073
	B			0.120	0.095
Mean values of θ_{st} (± S.E.) for all events				0.075 ±0.015	0.056 ± 0.004

* SO_4-S applied to paddock A and elemental S to paddock B

The fact that ΔS values calculated by eqn (11) matched measured values closely for the limited range of soil and environmental conditions examined so far implies that in an operational sense at least, a drained soil system subjected to episodic rainfall inputs appears to behave as a well mixed reactor in respect of solute leaching. The capacity of the reactor (the transport volume V_{st}) is, however, small relative to the total void volume of the soil. Given our knowledge of the complexity of coupled water and solute movement through structured soils (Bouma, 1981; Thomas and Phillips, 1979; White 1985b), elution through a well mixed system seems intuitively to be an unlikely physical model for solute leaching in a mole-drained soil. Thus, further field experiments should be done to determine the limits of soil, environmental and input conditions at which eqn (11) fails as a realistic descriptor of the complex transport system. In other words, the robustness of the concept of characterizing a soil by means of a space and time-averaged value of θ_{st} leading to predictions of solute losses by eqn (11), needs to be tested.

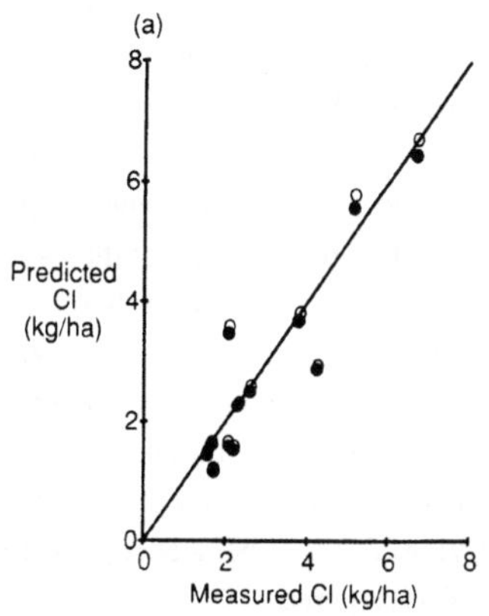

Figure 4a: A comparison of predicted and measured Cl leached.

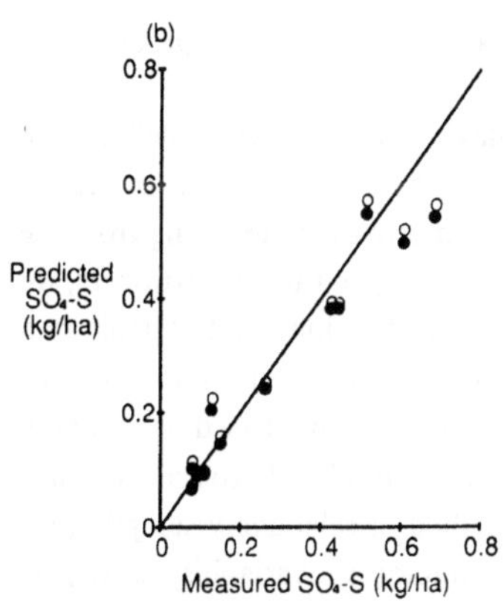

Figure 4b: A comparison of predicted and measured SO_4-S leached. Both figures use V_{st} derived from chloride transport in the drained Tokomaru soil (solid lines represent a 1:1 ratio).

Eqn (11) can be expanded in the form

$$\Delta S = C_i V_{st} (V_{ex}/V_{st} - (V_{ex}/V_{st})^2/2 + (V_{ex}/V_{st})^3/6 + ...) \qquad (12)$$

which shows that when V_{ex}/V_{st} is < 0.2,

$$\Delta S \approx C_i V_{ex} \qquad (13)$$

so that the quantity of solute leached should depend only on the initial concentration in the transport volume and the drainage volume. Alternatively, when V_{ex}/V_{st} is large (> 2.5),

$$\Delta S \approx C_i V_{st} \qquad (14)$$

None of the events at Wytham or Palmerston North approached the condition where eqn (14) could be used, and only 5 out of the 23 individual realizations of solute leaching at the two sites satisfied the condition for eqn (13) to be used. In the latter case, recent work by Scotter et al. (1990) in developing a simplified analysis of transient water flow to drains indicates that V_{ex} can be estimated from easily measured soil properties (saturated hydraulic conductivity, total porosity and macroporosity), drain spacing and depth, drainage coefficient, rainfall and evaporation. Irrespective, however, of whether V_{ex} is measured or estimated, for the majority of events at any one site an estimate of the space and time-averaged θ_{st} (giving V_{st}) will be needed; and for all events, an estimate of C_i for use in eqns (7), (11), (13) or (14) is required.

5. Estimation of solute concentrations in a soil's transport volume

In contrast to the reproducible transport characteristics of these drained grassland soils, as evidence by the reasonable stability of the space and time-averaged values of θ_{st}, the concentration of indigenous solutes in the transport volume appears to be highly variable (White, 1989). This is especially true of NO_3^-, and to a lesser extent SO_4^{2-}, both of which can be products or reactants in biological transformations.

Variability in solute concentration can occur at two scales: (1) on a "microscale" between regions of transport and non-transport soil volume, such as might exist between the outside and inside of large soil aggregates (White, 1985a); and (2) on a "macroscale" resulting from the spatially and temporally variable inputs of solute elements in dung and urine, fertilizers, through microbial activity (mineralization) and plant activity (nitrogen fixation), and variable removals or transformations through microbial activity (immobilization) and plant uptake. Macroscale variability can be accommodated to some extent by intensive sampling, based on recognition of the range of any spatially dependent variability (White et al., 1987). Microscale variability in structured soils is more difficult to cope with unless a method of selectively sampling solution from the soil's transport volume can be developed. Where microscale variability exists, as in the Evesham soil, measurements made on soil extracts in K_2SO_4 or KCl give average C_i values for the *whole* soil volume only, which have proved unsuitable for modelling NO_3^- leaching (White, 1989). Similar problems have been experienced with SO_4^{2-} measurements in the Tokomaru silt loam, and with NO_3^- measurements in the same soil (G.N. Magesan, unpublished).

To date, the best estimates of C_i for the soil's transport volume have been given by the solute concentration in the first volume of effluent that is collected (e.g. in the field measurements of SO_4^{2-} leaching - Table IV); or from the background concentration of NO_3^- in the drainage between events (Haigh and White, 1986). There is some evidence to suggest that SO_4^{2-} concentrations in the drainage from the Tokomaru soil treated with SO_4-S and elemental S do not change rapidly with cumulative drainage volume (L.K. Heng, unpublished), which if confirmed would simplify the prediction of SO_4-S leaching losses using eqn(11). However, the use of initial drainflow concentrations in the validation of the TFM as a predictive leaching model is unsatisfactory because the values are not independent of the data used to calibrate the model. In addressing the problem of estimating values of C_i for NO_3^-, a detailed study of spatial and temporal variation in the activity of nitrifying organisms in the Tokomaru soil has been made using a short-term nitrification assay (SNA) as a measure of nitrifier potential at any one site or time (Bramley and White, 1989). The SNA values fluctuated to some extent with soil moisture content and showed marked spatial dependence over a range of c. 3 m. Ways of extracting solution from a soil's transport volume only are also being explored. Resolution of this problem would be a major advance in the development of a predictive, stochastic model for the leaching of labile nutrient ions in field soils.

Acknowledgements

The authors are indebted to Dr D.R. Scotter, Dr N.S. Bolan, Ms V.O. Snow and Mr G.N. Magesan, colleagues and postgraduate students in the Department of Soil Science, for helpful discussions and advice.

References

Bouma, J., 1981: *Soil morphology and preferential flow along macropores*. Agricultural Water Management, **3**, 235-250.

Bramley, R.G.V. and R.E. White, 1989: *The effect of pH, liming, moisture and temperature on the activity of nitrifiers in a soil under pasture*. Australian Journal of Soil Research, **27** (in press).

Dyson, J.S. and R.E. White, 1987: *A comparison of the convection-dispersion equation and transfer function model for predicting chloride leaching through an undisturbed structured clay soil. Journal of Soil Science,* **38**, 157-172.

Haigh, R., A. and R.E. White, 1986: *Nitrate leaching from a small underdrained grassland clay catchment*. Soil Use and Management, **2**, 65-70.

Jury, W.A., 1982: *Simulation of solute transport using a transfer function model*. Water Resources Research, **18**, 363-368.

Jury, W.A., L.H. Stolzy, and P. Shouse, 1982: *A field test of the transfer function model for predicting solute transport*. Water Resources Research, **18**, 369-374.

Jury, W.A., G. Sposito and R.E. White, 1986: *A transfer function model of solute transport through soil. Fundamental concepts*. Water Resources Research, **22**, 243-247.

Lawes, J.B., J.H. Gilbert and R. Warrington, 1882: *On the amount and composition of the rain and drainage waters collected at Rothamsted*. William Clowes and Sons, London.

Nielsen, D.R., M.Th. van Genuchten and J.W. Biggar, 1986: *Water flow and solute transport processes in the unsaturated zone*. Water Resources Research, **22**, 89-108.

Rinaldo, A., A. Marani and A. Bellin, 1989: *On mass response functions*. Water Resources Research, **25**, 1603-1617.

Scotter, D.R., L.K. Heng, D.J. Horne and R.E. White, 1990: *A simplified analysis of soil water flow to a mole drain*. Journal of Soil Science, **41**, (in press).

Thomas, G.W. and R.E. Phillips, 1979: *Consequences of water movement in macropores*. Journal of Environmental Quality, **8**, 149-152.

Towner, G.D., 1989: *The application of classical physics transport theory to water movement in soil: development and deficiencies*. Journal of Soil Science, **40**, 251-260.

White, R.E., 1985a: *A model for nitrate leaching in undisturbed structured clay soil during unsteady flow*. Journal of Hydrology, **79**, 37-51.

White, R.E., 1985b: *The influence of macropores on the transport of dissolved and suspended matter through soil*. Advances in Soil Science, Vol. **3**, (Ed. B.A. Stewart: Springer-Verlag), 95-120.

White, R.E., 1987: *A transfer function model for the prediction of nitrate leaching under field conditions*. Journal of Hydrology, **92**, 207-222.

White, R.E., 1989: *Prediction of nitrate leaching from a structured clay soil using transfer functions derived from externally applied or indigenous solute fluxes*. Journal of Hydrology, **107**, 31-42.

White, R.E., J.S. Dyson, R.A. Haigh, W.A. Jury and G. Sposito, 1986: *A transfer function model of solute transport through soil. 2. Illustrative applications*. Water Resources Research, **22**, 248-254.

White, R.E., R.A. Haigh and J.H.Macduff, 1987: *Frequency distributions and spatially dependent variability of ammonium and nitrate concentrations in soil under grazed and ungrazed grassland*. Fertilizer Research, **11**, 193-208.

R.E. White, and L.K. Heng, Department of Soil Science, Massey University, Palmerston North, New Zealand

Field-Scale Water and
Solute Flux in Soils
Monte Verità

ANALYSIS OF CAISSON TRANSPORT EXPERIMENT BY TRAVEL TIME APPROACH

V. Nguyen, G. Dagan and E.P. Springer

The transport experiments conducted by Los Alamos National Laboratory for saturated flow in an intermediate-scale caisson are described. The results suggest that the solute spreading pattern has been dominated by the effect of spatial variability of hydraulic properties. A stochastic model of horizontally varying velocity distribution is employed in order to interpret the concentration data in terms of travel time distribution. A satisfactory match between the measured solute mass cross-sectional transfer in saturated flow and model is achieved for a standard deviation of the logconductivity of 1.2. The model may serve for future interpretation of transport in unsaturated flow and for reactive solutes conducted in the same caisson.

1. Introduction

Disposal of radioactive wastes in the unsaturated zone and their subsequent motion and decay calls for prediction with the aid of transport models. Usually, such models consist of a system of partial differential equations for the concentration C, function of space and time, which has to be solved with appropriate initial and boundary conditions. These equations are based on physical first principles, but contain a number of parameters that have to be determined experimentally. Hence, experiments are crucial for validating the equations and their solution and for determining the values of parameters. Traditionally the experimental work is carried out under laboratory conditions, e.g. in vertical porous columns in which water flow and solute transport are monitored under carefully controlled conditions. However, the question is whether such small columns, usually of a homogeneous nature, are representative of field conditions. Field experiments, the ultimate validation test, are costly and difficult to control and monitor. To cover this gap, the Los Alamos National Laboratory (LANL) has initiated and performed a series of experiments at an *intermediate* scale, in caissons much larger than laboratory columns, of depth comparable with natural soil, but of limited horizontal extent. These experiments, conducted with both conservative and reactive solutes, are described in a series of reports (Nyhan et al., 1986; Fuentes and Polzer, 1986; Polzer et al., 1986). Their main advantage is in expanding the scale on one hand, while permitting careful control and measurements on the other. Besides expanding the scale, the interest in the particular medium and its packing was related to the design of landfills in which the same material was emplaced under similar conditions. First attempts (Fuentes and Polzer, 1986; Polzer et al., 1986) to interpret the results by a best fit between concentration measurements and the solution of the one-dimensional transport equations with a constant dispersivity were not successful. Further attempts by a few teams, using more refined schemes (Springer and Fuentes, 1987), have not yielded models capable of satisfactory matching the measurements. These difficulties were attributed to the spatial variability of the medium properties, while the analyses were essentially deterministic. Under these conditions the number of concentration measurement stations, while sufficient in order to monitor a one-dimensional process, were not representative enough of a heterogeneous medium. Also, water and solute input conditions at the caisson surface were one-dimensional and uniform. To further elucidate the mechanisms underlying the transport process, another experiment was conducted in the same caisson under saturated flow conditions, with quadrupling the number of measurement stations (Fuentes et al., 1987) (see next Section). Again, attempts (Fuentes et al., 1987) to match the measured C and solution of the convection-dispersion equation with constant coefficients were not successful. To illustrate the point, the fitted dispersivity coefficient (see Table VII Fuentes et al., 1987), is represented in Fig. 4 as function of the conservative solute travel distance. It is seen that it displays extremely large values as compared to the pore-scale dispersivity, and grows considerably with distance. Therefore, even under the apparently uniform packing conditions of the caisson medium, spatial variability is present and has an important effect upon transport. The aim of the present study is to analyze and explain these findings by employing the recent developments of the stochastic theory of transport (see e.g. reviews Dagan, 1986; Gehlhar, 1986). More precisely, we wish to provide an adequate modeling framework, to validate it with the aid of the experiments and to assess the parameters characterizing the spatial variability of the hydraulic conductivity. This is carried out for saturated flow and the conservative solute. However, we plan to use the results in future for analyzing the behavior of reactive solutes and under unsaturated conditions. Some implications of the present work for these conditions are discussed in the last section here.

2. Experimental design

The caisson is a galvanized metal corrugated highway culvert stood upright. The dimensions of the caisson are 6 m high by 3 m diameter. Corrugated pipe was used to prevent channeling of flow along sidewalls. Also, the inner surface of the caisson was coated with teflon paint to minimize adsorption of solutes to the surface.

The design of the caisson facility used six experimental caissons around a central access caisson. Ports between the access caisson and the experimental caissons were located vertically on 75 cm centers (Figure 1) that allow horizontal placement of instruments. For the experiment reported in this study, the access ports were at depth z_i (i=1,2,3,4) of 36 cm, 113 cm, 264 cm, and 415 cm below the surface of the fill material, respectively. Besides, monitoring was conducted at the outlet pipe.

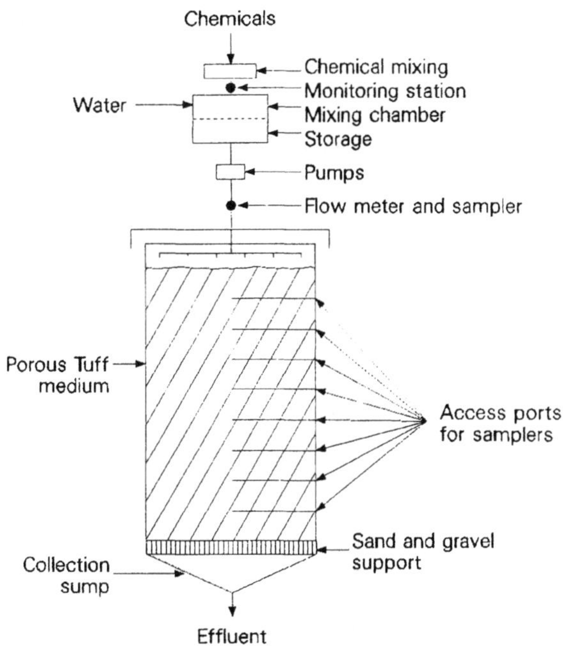

Figure 1: Schematic diagram schowing the caisson and its experimental features.

The sequence used in filling the caisson was an initial 0.25 m thick layer of gravel, another 0.25 m thick layer of coarse sand, and finally the backfill material of Bandelier Tuff (Figure 1, for details see Nyhan et al., 1986). The Bandelier Tuff was packed in 15 cm thick lifts and then compacted with a tamping machine to an approximate bulk density of 1.3 g/cm^3.
At each of the access ports, an aluminium neutron access tube was installed to the center of the caisson (150 cm). At each of four selected depths z_i four porous fiber solution samplers were installed. A probe was inserted in each quadrant, the coordinate x,y being drawn at random (Fuentes et al., 1987). The samplers were installed by coring the Banderlier Tuff and placing a tuff/silica flour slurry over the sampler and backfilling with crushed tuff.
The caisson was saturated by filling from the bottom with a 0.01 *N* (0.01 *M*) calcium chloride solution for a three-week period. Following saturation, 1 3 cm depth of the influent solution was maintained on the surface and a steady rate of Q/A=4 cm/d was established by regulating the

outflow. A six-day pulse of tracer was added beginning July 18, 1986. Samples from each of the four solution samplers at the four depths, as well as from the effluent were collected on a daily basis. Although strontium and iodide were added, this presentation will discuss only the iodide data.

3. Experimental results

Breakthrough curves for the four depths are presented in Figure 2. The response of the four samplers at the 36 cm depth is presented in Figure 2a. The curves are very close, and no iodide was detected after 16 days after initial application. Breakthrough curves at the largest depth of 415 cm (Figure 2d) display a more erratic response, and sampling was started after 10 days because a uniform travel time was assumed to prevail. While high concentrations were observed in the Q_1 and Q_4 samplers, the other two only breakthrough in the measurement period. The erratic behavior of the breakthrough curves at the 415 cm depth is displayed also at the other two sampling depths, 113 cm (Fig. 2b) and 264 cm (Fig. 2c). This behavior reveals the variability in flow paths in the horizontal plane across the caisson.

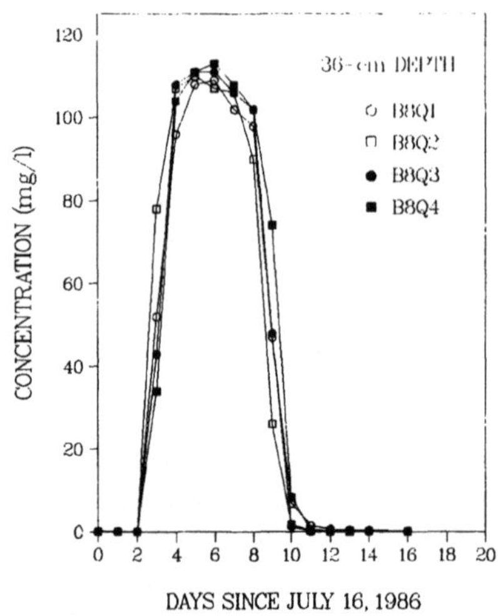

Figure 2a: The measured concentration as function of time at the four stations at depth: z_1*=36 cm*

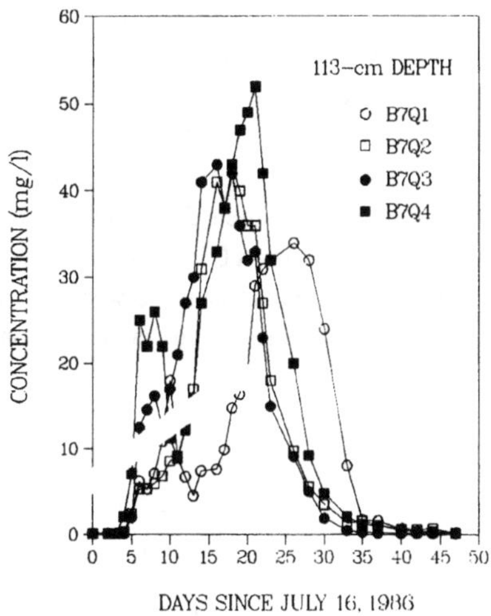

Figure 2b: The measured concentration as function of time at the four stations at depth: z_2=113 cm.

4. Qualitative analysis of transport and selection of model

The flow in the caisson was caused by a constant head difference applied at the two ends, with lateral no-flow conditions. Hence, the flow was steady, uniform and one-dimensional in the *average*. Let u(x,y,z) be the Darcian velocity at a point, z being a downward vertical coordinate with origin at the caisson upper boundary (Figure 1). Let $\bar{u}$ (z) = $(1/A)\int u \, dx \, dy = Q/\theta A$ be the average, constant, cross-sectional velocity, where A is the cross-section area, $\theta \cong 0.3$ is the water content (effective porosity) and Q is the water discharge. If the medium were homogeneous, $u \equiv \bar{u}$=const and correspondingly the streamlines would have been straight and vertical. In the actual, heterogeneous, medium, u = (K/θ)**J**, where **J** is the hydraulic gradient, is spatially variable due to spatial variability of the hydraulic conductivity K, and streamtubes are winding in space. Such streamtubes are represented qualitatively in Figure 3a. Furthermore, the velocities across any two streamtubes differ because each one presents a different resistance to flow, while the head drop is the same. The simplified model we have selected in order to represent this complex configuration is shown in Figure 3b: the actual streamtubes are replaced by a bundle of straight and parallel ones, such that both K and u= are functions of x,y solely. Hence, while the geometry is simplified, the velocity variation across streamtubes is preserved. This simplified model of flow and transport in unsaturated soil has been suggested by Dagan and Bresler (Dagan and Bresler, 1979) and can be shown to apply to general, three-dimensional, structures if the correlation scale in the z direction is larger than the caisson length. It can be shown then (Dagan, 1984) that the effective macrodispersivity grows linearly with travel distance. But this is precisely what is suggested by Figure 4, in which we have plotted the dispersivity λ=D/V, where D and V are best fits of the velocity and dispersion coefficients of Table VII (Fuentes et al., 1987), as function of travel distance. Once we adopt the model, the exact solution of flow through this stratified model is of constant head gradient J, equal to the head drop divided by the caisson length. Furthermore, for the assumed constant θ in saturated flow, u is proportional to K. In the next section we shall develop a simple relationship between u(x,y) and the solute mass flux (or the related flux averaged concentration) through any cross-section z.

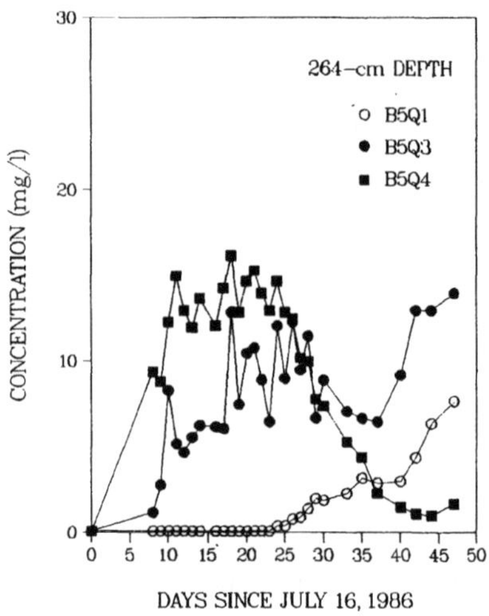

Figure 2c: The measured concentration as function of time at the four stations at depth: z_3= *264 cm.*

In the caisson experiment, neither K nor u were measured directly, but only $C(x_j,y_j,z_i,t)$, where the coordinates (j=4, i=6) of the measurements stations were fixed, whereas t, the time of sampling was running at short intervals (Figure 2). In order to compare the proposed model with measurements we have: (i) determined $M(z_i,t)$, where M is the mass of solute which has passed through the caisson cross-section $z=z_i$, by integrating the measured C, (ii) assessed the ability of the model to represent the measured M variation with z and t and (iii) inferred the statistical parameters characterizing the velocity distribution u(x,y) by a best fit of measured and theoretical results. From the latter the parameters of K distribution can be determined immediately, due to

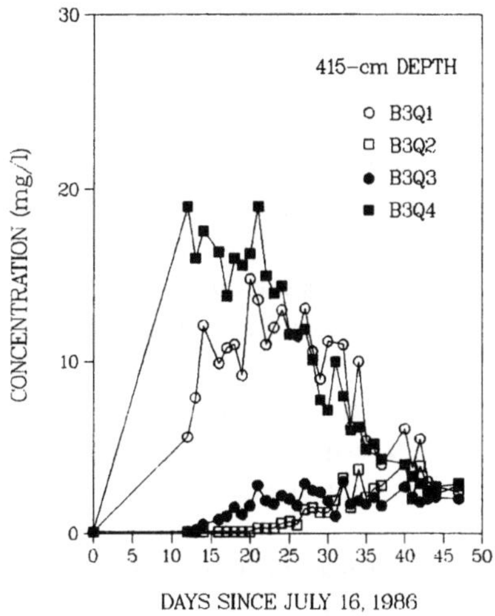

Figure 2d: The measured concentration as function of time at the four stations at depth: z_4=414 cm

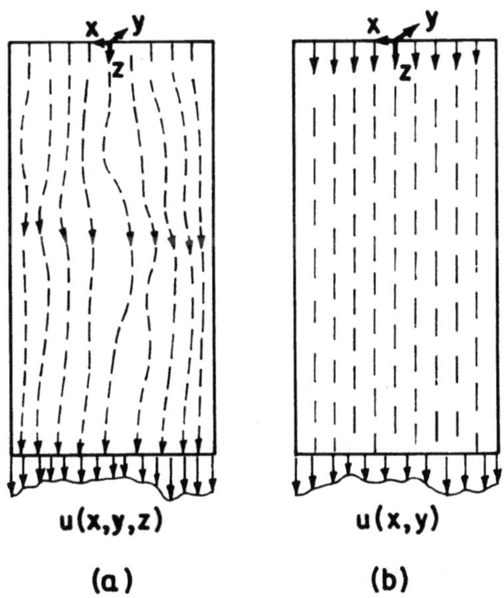

Figure 3: A sketch of streamtubes and cross-section velocity distribution: (a) actual (b) model.

the aforementioned proportionality. The application of this approach is described in the following sections.

5. The transport model (theory)

The velocity u(x,y) is presumably varying in an irregular manner in space, as suggested by the attempts to compute it at a few points (see Table VI Fuentes et al., 1987) from the measured C, which in turn varies quite erratically Figure 2). To reflect the uncertainty associated with this variation and in accordance with the stochastic approach (Dagan, 1986; Gelhar, 1986), u is regarded as a random space function. The simplest statistical characterization of u is by its p.d.f. (probability density function) f(u). Generally, f(u) changes in space, but u is assumed statistically *stationary,* so that f is a function of u solely. Furthermore, according to many available data (Freeze, 1975), K and subsequently u, may be regarded as lognormal. Hence, $\nu = \ln(u)$ is taken normal $N[m_\nu, \sigma_\nu]$ and the two fixed parameters $m_\nu = E[\nu]$ and $\sigma_\nu^2 = VAR[\nu]$ characterize completely ($f(\nu)$.

The solute mass flux through a cross-section at z is given by

$$Q_m = dM/dt = \theta \iint_A C(x,y,z,t)\, u(x,y,)\, dx\, dy \tag{1}$$

and subsequently the mass of solute which has crossed z at t is

$$M(z,t) = \theta \int_0^t \iint C(x,y,z,t)\, u(x,y)\, dx\, dy\, dt \tag{2}$$

We can relate M to u in a simple manner, since for our model it is easy to derive C in terms of u. This is done first for a pulse of solute of concentration C_0 inserted during the infinitesimal time $\Delta t'$ at $t=t'$, $z=0$. According to our model and with neglect of pore-scale dispersion, this pulse translates at constant velocity u along z. Hence, the concentration field is given by

$$\Delta C(x,y,z,t) = C_0\, \Delta t'\, u\, \delta[z-(t-t')u] \tag{3}$$

where δ is the Dirac distribution. Hence, the mass flux (1) for a pulse becomes

$$\Delta Q_m(z,t;t') = \theta\, C_0\, \Delta t' \iint_A u^2\, \delta[z-(t-t')u]\, dx\, dy \tag{4}$$

and ΔQ_m is random through its dependence on u. Its expected values is given by

$$\langle \Delta Q_m \rangle = \theta\, C_0\, \Delta t' \iint_A \{\int u^2\, \delta[z-(t-t')u]\, f(\nu)\, d\nu\,]\} dx\, dy = \theta\, C_0\, A\, z\, \Delta t'\, (t-t')^{-2} f[\ln z/(t-t')] \tag{5}$$

since the term in brackets is independent of x,y.

Substitution of the lognormal distribution for f, integration over t' for a finite pulse of constant C_0 applied at $z=0$ in the interval $0<t'<t_0$ and additional integration over time to obtain M, leads to the final results

$$\frac{\langle M(z,t)\rangle}{\langle M_0\rangle} = \frac{t}{2} - \frac{z}{2} + \frac{t}{2}\,\mathrm{erf}(A) - \frac{z}{2}\,\mathrm{erf}\left(B - \frac{\sigma_v}{\sqrt{2}}\right) \quad (0<t<1)$$

$$\frac{\langle M(z,t)\rangle}{\langle M_0\rangle} = \frac{1}{2} - \frac{t-1}{2}\,\mathrm{erf}(B) + \frac{z}{2}\,\mathrm{erf}\left(B - \frac{\sigma_v}{\sqrt{2}}\right) + \frac{t}{2}\,\mathrm{erf}(A) - \frac{z}{2}\,\mathrm{erf}\left(A - \frac{\sigma_v}{\sqrt{2}}\right) \quad (t>1) \qquad (6)$$

where dimensionless variables are defined by $t = t/t_0$, $z = z/\langle u\rangle t_0$, whereas $\langle M_0\rangle = \theta C_0\langle u\rangle A t_0 = C_0 Q t_0$ is the total solute mass introduced in the caisson. The functions A and B are given by $[\ln(t/z)+\sigma_v^2/2]/(\sigma_v\sqrt{2})$ and $\{\ln[(t-1)/z]+\sigma_v^2/2\}/(\sigma_v\sqrt{2})$, respectively. This completes the theoretical developments of our model that relate the expected value of the solute mass across z at t to the parameters $\langle u\rangle$ and σ_v which characterize the p.d.f. of the velocity u. This solution simplifies for z >1, tending to that corresponding to a concentrated pulse of mass M_0 inserted at $t=t_0/2$ at z=0

$$\frac{\langle M(z,t)\rangle}{\langle M_0\rangle} = \frac{1}{2}[1 + \mathrm{erf}(C)] \quad ; \quad C = \ln\left(\frac{\tau+\sigma_v^2/2}{\sigma_v\sqrt{2}}\right) \quad ; \quad \tau = (t-0.5)/z \qquad (7)$$

Differentiation of $\langle M\rangle$ with respect to t leads to $\langle Q_m\rangle$ which in turn is related to the flux averaged concentration $\langle C_f\rangle = \langle Q_m\rangle/\theta\langle u\rangle A$. We adhere here to M as the basic function to connect concentration with velocity statistics, the latter being proportional to hydraulic conductivity for the present model.

By a similar procedure one may compute VAR(M) and even higher statistical moments of M. They depend, however, on the spatial correlation of u, i.e. on the covariance of u at two points in the plane. We do not pursue such computations here, but adopt the basic assumption that $\mathrm{VAR}(M)/\langle M\rangle <<1$. This is equivalent to assuming ergodicity, i.e. $\langle M\rangle$ is approximately equal to M in any realization. In simple words, we assume that in the given caisson experiment the entire population of u values is spanned by the fluid. This assumption may be checked by inferring the integral scale of u, and by verifying that it is smaller the caisson diameter. This avenue is not followed here, but an indirect support is found from the agreement between the theoretical results for $\langle M\rangle$ and measurements. The great advantage of employing M as the basic variable is precisely in its lesser variability, whereas C itself is much more fluctuating. In other words, although our model is not able by its vary nature to reproduce C at a given point, it may predict some spatially averaged variables like M.

We are now in a position to compare the theoretical $\langle M(z,t)\rangle$ (6,78) with the measured one.

6. Determining M(z,t) from the measured concentration

The basic data are the concentration measurements $C(x_j,y_j,z_i,t_k)$ (j=1,..,4 ; i=0,...,5), where the time has been discretized at 1 day intervals (k=1,...,45). Our next step was to use polynominal interpolation (Press et al., 1987) over time at each station and subsequently over space at each t_k

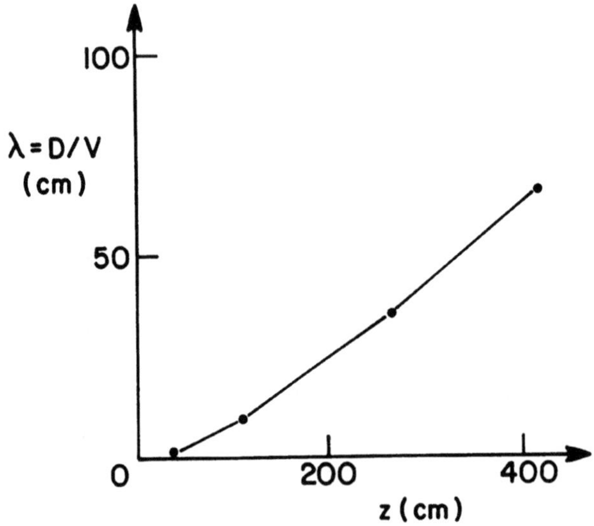

Figure 4: The dependence of the effective dispersivity on the distance from the inlet. Based on values of dispersion coefficient and velocity of Table VII (Fuentes et al., 1987).

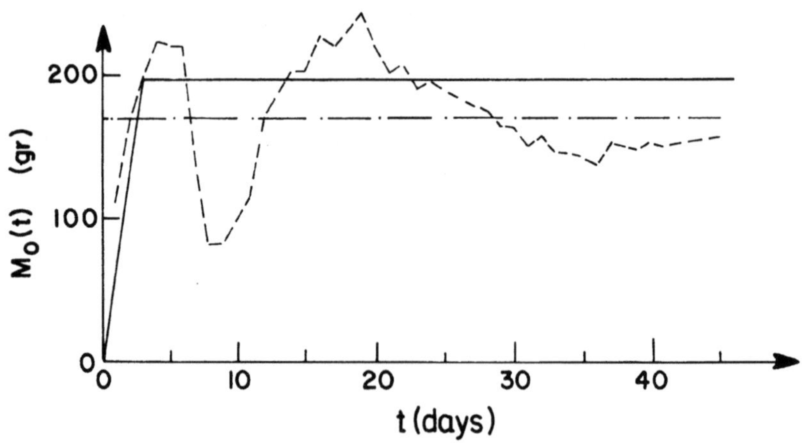

Figure 5: The dependence of the solute mass which has passed the inlet section z=0 upon time (full line-Mot based on measured C_0 and Q in the inlet reservoir, dashed line-M_0 determined by temporal and spatial interpolation of measured concentrations values, dashed-dotted line- the average M_0).

in order to reproduce C everywhere. The measured C values were attributed to the entire quadrant in the x,y plane to which x_j,y_j belonged. Subsequently, $M(z_i,t_k)$ has been computed by integrating C over the caisson form z_1 to z_6 and adding to it the mass outflown through the outlet. The latter is equal to the outlet concentration $C(z_5,t)$ multiplied by Q and integrated over time from t=0 to $t=t_k$. $M(x,y,0,t)=M_{0t}(t)$ is the actual input solute mass, which grows linearly with t for $0<t<t_0$ and stays subsequently constant and equal to $M_{0t}=C_0t_0Q \geq 197$ gr. In Figure 5 we have depicted the variation of M_0 with t, determined by spatial interpolation, as well as M_{0t}, the theoretical mass. It is seen that M_0 does not coincide with M_{0t}, but varies around it. This variation, which is due to the limited number of concentration measurements, to its erratic variation, to measurement errors and to interpolation errors, is not as large as suggested by Figure 5.

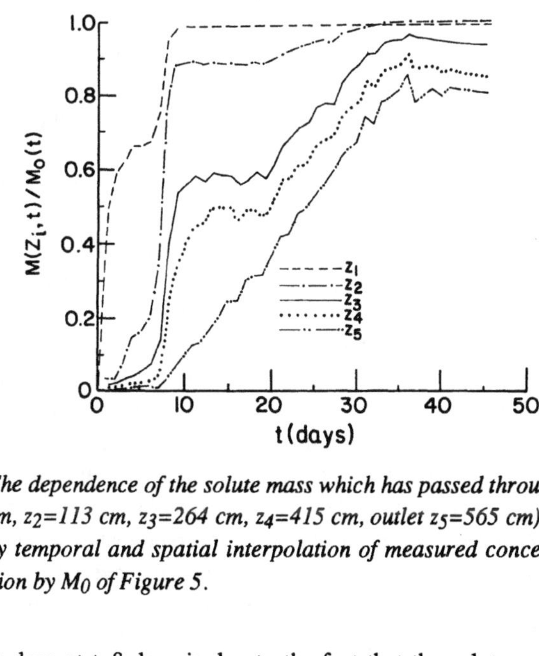

Figure 6: The dependence of the solute mass which has passed through cross-section z_i (z_1=36 cm, z_2=113 cm, z_3=264 cm, z_4=415 cm, outlet z_5=565 cm) upon time, determined by temporal and spatial interpolation of measured concentration values. Normalization by M_0 of Figure 5.

Indeed, the large drop at t=8 days is due to the fact that the solute pulse was between two measurement levels in that period. The drop with time for t>20 days reflects the fact that by that time most of the mass flux is at the outlet. For some unexplained reason it seems that the outlet total mass up to t=45 days does not approach the theoretical limit M_{0t}. Yet, the average M_0 over time for the entire period is equal to 170 gr, which can be considered quite satisfactory. The uncertainty associated with M_0 can be incorporated in the analysis, but this topic is not pursued here. Instead, in order to compare measurements with the theoretical result, we have normalized the measured $M(z_i,t)$ with respect to $M_0(t)$, for $t>t_0$. This can be considered as a conditioning of M(z,t) with respect to the total one determined by using the C measurements, which is affected by errors, rather than to the fixed total mass which has crossed the inlet M_{0t}. However, we have carried out similar computations for $M(z_i,t)/M_{0t}$, with results close to the one presented here. The curves $M(z_i,t)/M_0$ are reproduced in Figure 6 for the five stations z_i (i=1,...,5) and although the curves still display some fluctuating behavior, they are much less erratic than the original C measurements, supporting our claim about the smoothing effect of integration over the cross-section and depth. These curves are considered here as the experimental data to be used for farther analysis and comparison with the theoretical result (Eqs. 6,7).

7. Comparison between measurements and model

The theoretically determined $\langle M\rangle/\langle M_0\rangle$ (Eq. 6) depends on the two parameters which characterize the velocity distribution, namely $\langle u\rangle=\exp(m_v+\sigma_v^2)$ and σ_v. However, the value of $\bar{u}$ is known, since both Q, the water discharge, and θ, the porosity, have been measured independently. Hence, the corresponding value of $\langle u\rangle=\bar{u}=14$ cm/day has been taken as given. The next step was to determine the hintherto unknown σ_v by a best fit between the theoretical (Eq. 6) and measured (Figure 6) M. The degree of agreement between the two for a fixed σ_v is considered to be a criterion of validity of the model. Of course, a definitive test would have been achieved by a separate measurement of hydraulic conductivity of a sufficiently large number of undisturbed samples, but this was not available. By substituting $z=z_i$ (i=l,..,5) in Eq. 6 and by a least square procedure applied to the theoretical $\langle M\rangle/\langle M_0\rangle$ at various z_i and t and measurements, a best fit has been achieved for σ_v=1.2. For this value the ratio between the average deviation and the mean $\langle M\rangle/\langle M_0\rangle$=0.5 has been found to be equal to 0.18. To illustrate the degree of agreement, as well as the soundness of the assumed model, we have represented in Figure 7 the measured M/M_0 of Figure 6, but this time in terms of the dimensionless variable τ defined in Eq. 7 and for the stations z_i (i=2,...,5). It should be remembered that for these stations, satisfying the requirement $z>1$, the theoretical model (6) degenerates into (7), which stipulates that $\langle M\rangle/\langle M_0\rangle$ is function of the self-similar variable τ, rather than z and t separately. Examination of Figure 7 indeed shows that the measured M/M_0 indeed become quite close in this representation, although they do not collapse on a unique curve, for the various reasons enumerated above. Also in Figure 7 we have represented the theoretical curve (Eq. 7) for σ_v= 1.2 (full line) as well as two bounding curves corresponding to σ_v=0.68 and 1.72, respectively. These two values correspond to an interval of confidence of 98% derived from the least squares procedure. In view of the various errors affecting measurement and interpolation procedures, the agreement can be considered quite satisfactory.

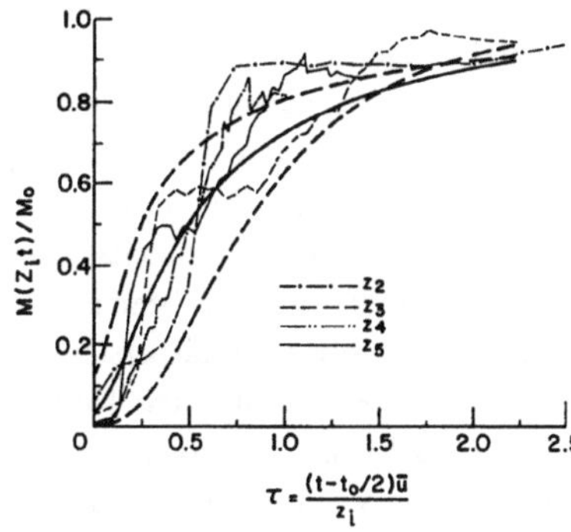

Figure 7: Representation of M/M_0 of Fig. 6 (thin lines) as functions of the self-similar variable τ. The theoretical solution (Eq. 7) for σ_v=1.2 (thick full-line) and for the interval of confidence valueσ σ_v=0.68 and σ_v=1.78 (thick dashed lines).

8. Travel time distribution

The expected values of the solute mass flux $<\Delta Q_m(z,t;t'>$ (4) for a slug of solute is proportional to the probability density function g(T,z;t') of the travel time T. The latter is defined as the time at which a solute particle entering the caisson at z=0, t=t′ crosses the plane at z. The equivalence between the flux and g has been established in previous studies of representation of transport by travel (transfer, arrival) time (Jury et al., 1986; Shapiro and Cvetkovic, 1988; Dagan and Nguyen, 1989). Consequently, the function $<M(z,t>/M_0$ (7) and Fig. 7, for a slug inserted at $t=t_0/2$ at z=0, represents the probability distribution function $G/T,z;t_0/2)$ where G(T,z;t′) = ∫g (T'', z; t') dT''. The connection between ⟨M⟩ for a pulse of finite width and G for a slug (Dagan and Nguyen, 1989) is not discussed here. The dependence of G upon the similarity variable *t* of Eq. (7) has been suggested in the past (Jury et al., 1986), but here it has been supported by a physical model (Dagan and Bresler, 1979) and applied to a controlled experiment. In this sense the present study constitutes one of the first attempts to develop a comprehensive stochastic travel time model and to apply it to actual data.

9. Summary and conclusions

The present study presents a model of the caisson transport experiments of a conservative solute in saturated flow. These experiments have the unique feature of being carried out at a much larger scale than the usual laboratory columns, but under carefully controlled conditions, seldom attained in the field. The analysis of the concentration measurements carried in the past (Fuentes et al., 1987; Springer and Fuentes, 1987) indicated that the medium is heterogeneous and the transport is strongly affected by spatial variability. This point is underscored by Figure 2 and Figure 4, which displays the dependence of an apparent effective dispersivity λ upon the distance z from the inlet. The large values of λ compared to pore-scale dispersivity indicate indeed that this is a "macrodispersion" effect. Furthermore, the linear growth with z suggests that a stochastic model with velocity varying mainly in the horizontal plane x,y, is appropriate. A simple model (Dagan and Bresler, 1979) of this nature has been adopted, by assuming that u is lognormally distributed. This model neglects pore-scale dispersion and the solute rectangular pulse introduced at the inlet, travels along the vertical unaltered. However, the velocity differs in the horizontal plane. This simplified model cannot reproduce the complex variation of the measured C, but it is assumed to provide a valid representation of spatially averaged concentration or its statistical moments. Such an average quantity that we have chosen as representative is the solute mass M past a given caisson cross-section at z, as function of time. It is related to the mass flux or to the cross-sectional flux averaged concentration. This quantity has been evaluated from the concentration measurements by interpolation and integration of the observed data. Due to various errors (finite sample, measurement errors, interpolation) the total mass flowing through the inlet section M_0 did not match M_{0t}, the one based on the initial concentration in the inlet reservoir. After filtering out this effect, a best fit between the measured and theoretical M/M_0 could be achieved in a satisfactory manner after adopting an ergodic hypothesis for M/M_0, i.e. assuming that $\langle M\rangle \cong M$. This has led to the value of the standard deviation $\sigma_V = 1.2$ of the logvelocity. Due to the linear dependence, this is also the inferred logconductivity standard deviation.

The present analysis can be related to the representation of transport by travel time and all results can be converted into travel time probabilities. Hence, this representation has a definite potential for analysis of solute motion in the soil.

The analysis can be refined in a few respects, such as better interpolation procedure, statistical analysis of the variation of M_0, filtering out of measurement errors at certain t, identification of the correlation structure of u and development of more complex models. Whether the amount and quality of data justify such an effort is a matter of future consideration. Other subjects of definite interest that we plan to study are: (i) the implication of the present findings upon strategy of data collection in future experiments and field studies and (ii) the use of the present results for analyzing the experiments on unsaturated flow and reactive solutes.

In conclusion, the stochastic theory of transport in heterogeneous media has proved to be a valuable tool for explaining and interpreting the caisson saturated flow experiments and a promising approach for future similar investigations.

References

Dagan, G., 1986: *Statistical theory of groundwater flow and transport: pore to laboratory, laboratory to formation and formation to regional scale.* Water Resour. Res., **22**, 120S-135S.

Dagan, G, and E. Bresler, 1979: *Solute dispersion in unsaturated heterogeneous soil at field scale I: Theory.* Soil Science Soc. of Am. Journ. **43**, 461-467.

Dagan, G., 1984: *Solute transport in heterogeneous porous formations.* J. Fluid Mech. **145**, 151-177.

Dagan, G., and V. Nguyen, 1989: *A comparison of travel time and concentration approaches to modeling transport by groundwater.* Journ. Contaminant Hydrology, **4**, 79-91.

Freeze, R.A., 1975: *A stochastic-conceptual analysis of one-dimensional groundwater flow in nonuniform homogeneous media.* Water Resour.Res., **11**, 725-741.

Fuentes, H.R., and W.L. Polzer, 1986: *Interpretative analysis of data for solute transport in the unsaturated zone.* NUREG/CR-4737, LA-10817-MS, Los Alamos National Laboratory.

Fuentes, H.R., W.L. Polzer and E.P. Springer, 1987: *Effects from influent boundary conditions on tracer migration and spatial variability features in intermediate-scale experiments.* NUREG/CR-4901, LA-10981-MS, Los Alamos National Laboratory.

Gelhar, L.J., 1986: *Stochastic subsurface hydrology from theory to applications.* Water Resour. Res., **22**, 135S-145S.

Jury, W.A., G. Sposito, and R.E. White, 1986: *A transfer function model of solute transport through soil. 1, Fundamental concepts.* Water Resour. Res., **22**, 243-247.

Nyhan, J. et al., 1986: *A joint DOE/NRC field study of tracer migration in the unsaturated zone.* LA-10575-MS, UC-708, Los alamos National Laboratory.

Polzer, W.L., E.H. Essington, H.R. Fuentes, and J.W. Nyhan, 1986: *Compilation of field-scale caisson data on solute transport in the unsaturated zone.* NUREG/CR-4720, LA-10798-MS, Los alamos National Laboratory.

Press, W.H. et al., 1987: *Numerical recipes- the art of scientific computing.* Cambridge Univ. Press.

Shapiro, A.M., and V.D. Cvetkovic, 1988: *Stochastic analysis of solute arrival time in heterogeneous porous media.* Water Resour. Res., **24**, 1711-1718.

Springer, E.P. and H.R. Fuentes (eds), 1987: *Modeling study of solute transport in the unsaturated zone.* Workshop Proceedings, NUREG/CR-4615, LA-10730-MS, Vol.2, Los Alamos National Laboratory.

V. Nguyen, EWA, Inc., 133 First Avenue North, Minneapolis, MN, USA

G. Dagan, Fac. of Engnr., Tel Aviv University, Tel Aviv, Israel

E. P. Springer, Los Alamos National Lab., Los Alamos, NM, USA

Field-Scale Water and
Solute Flux in Soils
Monte Verità

FIELD ESTIMATES OF HYDRAULIC CONDUCTIVITY FROM UNCONFINED INFILTRATION MEASUREMENTS

M. D. Ankeny*, R. Horton, and T. C. Kaspar[1]

Soil water infiltration exhibits spatial variability due to both intrinsic soil properties and management effects (e.g. wheel traffic and tillage). The description or prediction of field scale water and solute movement, therefore, requires that the spatial distribution of hydraulic properties be known. Practical measurement of hydraulic properties of unsaturated soil requires both a rapid field technique to sample the necessary number of sites and a straightforward analytical technique for interpreting data. We have designed precision automated tension infiltrometers that enable rapid determination of soil water infiltration. Tension infiltrometers were used to obtain unsaturated infiltration data for chisel-plow, no-till, wheel-trafficked, and non-wheel trafficked management zones. This paper reports means and coefficients of variation for hydraulic conductivities at multiple tensions derived from unconfined infiltration measurements. Mean hydraulic conductivities decreased as tension increased. Trafficked sites had lower values of hydraulic conductivity than non-trafficked sites. The tension infiltrometer is a useful tool for obtaining spatial measurements of field infiltration and unsaturated hydraulic conductivities.

1. INTRODUCTION

Although knowledge of soil unsaturated hydraulic conductivities is important in many facets of soil science, adequate data to characterize field hydraulic conductivities are difficult to obtain. Saturated and near-saturated hydraulic properties are of particular interest and are difficult to predict independently of measurements. This interest recently led several groups to improved methods for measuring infiltration under tension (Elrick et al., 1988a; Ankeny et al., 1988; Perroux and White, 1988). The objective of this paper is to show the application of tension infil-

[1]National Soil Tilth Laboratory, USDA-Agricultural Research Service and Dept. of Agronomy, Iowa State Univ., Ames, Iowa, 50011. Joint contribution from USDA-ARS and Iowa State Univ. Journal Paper No. J-13836 of the Agric. and Home Econ. Ext. St., Ames, IA. Projects No. 2659 and 2715. Received October, 1989. *Corresponding author.

tration equipment, techniques, and analysis to the examination of field variability of hydraulic properties.

2. MATERIALS AND METHODS

A modified tension infiltrometer (Fig. 1) was designed to improve infiltration-rate measurement precision and tension control as well as to automate data collection and analysis (Ankeny et al., 1988). Precision of infiltration-rate measurement is approximately tenfold better than manual measurement, and a single worker can operate several devices concurrently. A similar device (in essence, an automated single-ring infiltrometer) is used to measure saturated infiltration rates.

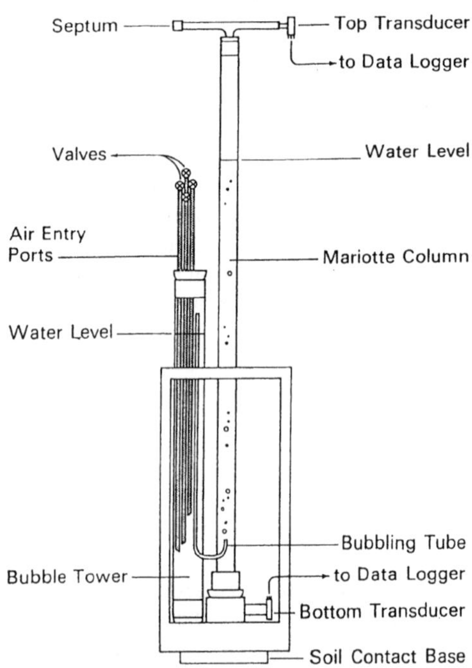

Figure 1: Schematic diagram of tension infiltrometer.

Tillage and controlled traffic treatments were established on a Tama soil (a fine-silty, mixed, mesic typic Argiudoll) 60 km east of Ames, IA, USA. Corn (Zea Mays L.) and soybeans (Glycine max L. Merr.) were grown in rotation for four years before infiltration measurements. Infiltration rates were measured on trafficked and non-trafficked interrows in chisel-plow and in no-till treatments. No-till plots received no primary tillage, but were cultivated once a year. Chisel plow plots were fall chiseled, disked shortly before planting, and cultivated once a year. Unconfined infiltration-rate measurements were made at four surface water conditions, starting with ponded infiltration and proceeding to flow rate measurements at 30, 60, and 150 mm of water tension (negative pressure). A sharpened 7.62 cm diameter ring was pressed about 1 cm into the soil to limit the infiltration area. After several minutes of infiltration at each tension, data were recorded for 1000 seconds at the steady-state unconfined infiltration rates. Eight sites were measured for each tillage-traffic combination.

3. THEORY

We use the following method based on Wooding's work (1968) to calculate hydraulic conductivities from unconfined infiltration data. Wooding proposed the following algebraic approximation of steady-state unconfined saturated infiltration rates into soil from a circular source of radius r

$$Q = \pi r^2 K + 4r\phi \tag{1}$$

where Q is the water flux, K is the saturated hydraulic conductivity and ϕ_m is a matric flux potential where

$$\phi(0) = \int_{\psi_i}^{0} K(\psi)\, d\psi \tag{2}$$

Rewriting equation 2 for a general infiltrating boundary condition, ψ, where ψ_i is the initial water potential of the unwetted soil yields

$$\phi(\psi) = \int_{\psi_i}^{\psi} K(\psi)\, d\psi \qquad , \psi > \psi_i \tag{3}$$

Measuring steady infiltrating fluxes $Q(\psi_1)$ and $Q(\psi_2)$ at two water potentials yields two equations and four unknowns

$$Q(\psi_1) = \pi r^2 K(\psi_1) + 4r\,\phi(\psi_1) \tag{4}$$

$$Q(\psi_2) = \pi r^2 K(\psi_2) + 4r\,\phi(\psi_2) \tag{5}$$

A third equation is obtained by assuming a constant K/ϕ ratio. Fig. 2 (after Elrick et al., 1988b) shows the basis for a numerical approximation that yields the fourth equation. The difference between $\phi(\psi_1)$ and $\phi(\psi_2)$ is approximately $\Phi(\Psi_1)-\Phi(\Psi_2) = \Delta\Psi[K(\Psi_1)+K(\Psi_2)]/2$.

We can now calculate hydraulic conductivity from any pair of unconfined infiltration rates taken at different tensions.

We estimated the hydraulic conductivity $K(\psi)$ at tension ψ calculated from the (ψ_1, ψ_2) rate pair (shown as subscripts) as

$$\begin{aligned} K_{sat} &= K(0) = K(0)_{0.30} \\ K(30) &= \frac{1}{2}\left[K(30)_{0.30} + K(30)_{30.60}\right] \\ K(60) &= \frac{1}{2}\left[K(60)_{30.60} + K(60)_{60.150}\right] \\ K(150) &= K(150)_{60.150} \end{aligned} \tag{6}$$

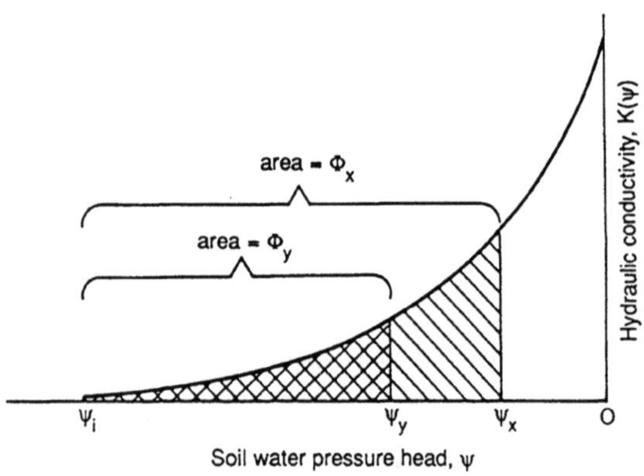

Figure 2: A generalized soil water pressure head-hydraulic conductivity relationship

4. RESULTS

Fig. 3 shows the effect of tillage and wheel traffic on infiltration rates at different nominal pore diameters (diameter is inversely proportional to tension). As sequentially smaller macropores emptied, the infiltration rate decreased for all treatments. As expected, traffic reduced infiltration rates in both tillages at all tensions. In addition, traffic changed the slope of the lines in Fig. 3.

Traffic significantly decreased infiltration at all tensions and trafficked no-till interrows were significantly higher infiltration rates than trafficked chisel-plow interrows. A decrease in intercept with constant slope would show proportional reduction in pore size classes. A decrease in slope shows flow in larger pores is more affected by compaction than flow in smaller pores. Our interpretation is that traffic compaction destroys more large than small macropores.

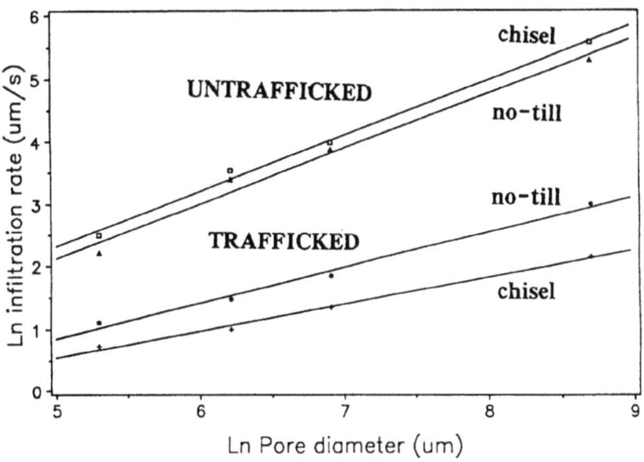

Figure 3: Plot of the log of the unconfined infiltration rate versus largest nominal water-filled pore diameter.

Table I. shows field variability found when unconfined hydraulic conductivity is examined by individual tensions for each treatment. Coefficients of variation ranged from 17% to 103% with a value around 50% being typical. The tension infiltrometers have operated and functioned well on a wide range of soil types (loess and glacial-till parent materials) and tillage management systems (no-till and chisel-plowed). Errors in estimating the steady-state infiltration rates were typically 1-5%. Thus we expect tension infiltration measurements to allow improved modeling efforts because saturated and unsaturated data can be obtained at each sample site and data can be recorded rapidly.

SUMMARY

Field scale predictions of water and solute movement generally require information describing distributions of soil hydraulic properties. Improved equipment, field techniques, and analysis now make it feasible to make accurate measurements of the distribution of unconfined infiltration and/or hydraulic conductivity. The improved description of individual sites (multiple measurements per site), as well as the feasibility of measuring many sites, makes these tools useful in obtaining needed information about field hydraulic conductivity and its variability.

Table I. Summary of means and coefficients of variation for estimated hydraulic conductivity of a Tama silty clay loam.

Treatment	Tension (mm H_2O)	Hydraulic Conductivity m/s	CV %
non-trafficked no-till	0	128.9	53
	30	23.0	50
	60	9.1	43
	150	2.5	35
Trafficked no-till	0	11.6	61
	30	2.7	38
	60	1.0	50
	150	0.3	68
Non-trafficked chisel	0	169.0	50
	30	23.2	31
	60	9.0	17
	150	2.9	20
Trafficked chisel	0	3.7	103
	30	1.7	53
	60	0.6	36
	150	0.2	68

References

Ankeny, M. D., T. C. Kaspar, and R. Horton. 1988: *Design for an automated tension infiltrometer*. Soil Sci. Soc. Am. J., **52**. 893-896.

Elrick, D. E., W. D. Reynolds, N. Baumgartner, K. A. Tan, and K. L. Bradshaw, 1988a: *In-situ measurements of hydraulic properties of soils using the Guelph permeameter and the Guelph infiltrometer*. In Proc. Third Int. Workshop on Land Drainage, Columbus, Ohio, Dec. 7-11, 1987.

Elrick, D. E., W. D. Reynolds, and K. A. Tan., 1988b: *A new analysis for the constant head well permeameter*. p. 88-95. In P. J. Wierenga (ed.) Proc. Int. Conf. and Workshop on the validation of flow and transport models for the unsaturated zone. Ruidoso, NM. May 22-25, 1988. New Mexico State University, Las Cruces, NM.

Perroux, K. M., and I. White, 1988: *Designs for disc permeameters*. Soil Sci. Soc. Am. J., **52**. 1205-1215.

Wooding, R. A., 1968: *Steady infiltration from a shallow circular pond*. Water Resour. Res., **4**, 1259-1273.

Mark D. Ankeny, R. Horton, and T.C. Kaspar, National Soil Tilth Lab., 2150 Pammel Drive, Ames, IA, 50010, USA.

Field-Scale Water and
Solute Flux in Soils
Monte Verità

SPRINKLER IRRIGATION, ROOTS AND THE UPTAKE OF WATER

B. E. Clothier, K. R. J. Smettem, and P. Rahardjo

Clearer perception and better modelling of soil water and chemical flow will be aided by improved measurement of the appropriate saturated and unsaturated flow characteristics of field soil. *In situ* measurements, at and near saturation are presented for two contrasting soils: the herbicided, coarse soil of an apple orchard, and a biologically-active, finer-textured soil growing pasture. These data are combined with more-unsaturated data from laboratory analysis on undisturbed cores. Soil water content measurements made after sprinkler irrigation of the orchard are interpreted in terms of simple notions of unsaturated flux-infiltration. Our water extraction observations stress the prime role of surface roots. These deep-rooted apple trees exhibited a depth-wise flexibility in root water uptake.

1. Introduction

Despite recent theoretical advances and huge increases in the ability merely to simulate larger-scale soil water transport, it still remains very difficult to observe infiltration in the field, and to characterise appropriately the absorption and transmission characteristics of rootzone soil.

Description of water movement in field soil is bedevilled by the partitioning of flow between the interconnected domains of the macropore system, and the matrix. Sufficiently free-water at the orifice of macropores can result in rapid and often chaotic movement of water and solute through the soil. By comparison, the flow of water through the matrix tends to more pedestrian. Measurements of the often-fragile, near-saturated physical properties of field soil should be performed *in situ*, and be capable of easy replication in both time and space.

Determination of the soil's unsaturated wetting characteristics is difficult *in situ*. But since the structure of the matrix tends to be quite robust, it seems appropriate to measure these in the laboratory. We combine the laboratory wetting unsaturated hydraulic conductivities with field measurements near to, and at saturation.

Plants, by virtue of their root uptake of water can play a dominant role in determining the magnitude, and even the direction of field soil water flow and solute transport. Whereas we currently have some well-established physico-chemical prejudices concerning water and solute flow in sterile, but reactive porous media, our understanding of water and nutrient transport in the presence of roots has less rigour. Indeed development of such physico-biological principles is somewhat restricted by a dearth of field information on the uptake behaviour of complete root systems. We interpret our spatially-integrated measurements of the total rate of soil water extraction by an apple tree, and compare them with the rate of apple tree water use determined in the trunk by the heat-pulse sap-flow technique.

2. A Matrix-Macropore Dichotomy.

To measure *in situ* the near-saturated hydraulic character of two field soils, we employed disc permeameters (Perroux and White, 1988) of at least two contrasting radii. From their respective steady flow we deduced the sorptivity S, and the hydraulic conductivity K (Smettem and Clothier, 1989). The prescribed surface pressure potential, ψ_o, ranged from -100 mm up to saturation. For drier potentials, down to about -0.9 m, the transient inflow/outflow technique of Scotter and Clothier (1983) was performed in the laboratory on 50 mm wide undisturbed cores of length 50 mm. Two soils were studied. The Manawatu fine sandy loam was under ryegrass-clover pasture and regularly grazed by dairy cows. Measurements on the Twyford sandy loam were conducted within the herbicided strip along a line of apple trees in a commercial orchard. The combined laboratory and field conductivity data are presented in Fig. (1) [after Clothier and Smettem, 1990]. Whereas the two soils have different matrix permeabilities, their saturated conductivities ($K_s = K[\psi_o=0]$) are, within observation error, identical. The slope of the unsaturated $K(\psi)$'s could have been guessed. Unsaturated water movement imposed by $\psi_o < -35$ mm into the finer-textured Manawatu soil would be characterised by a flow-weighted 'mean' pore-size of about 20 μm (Perroux and White, 1988). So capillarity is more dominant here than during similarly-unsaturated flow in the coarser Twyford sandy loam, where the equivalent pore size is about 50 μm.

The saturated conductivity need not however be related to either the soil's particle size distribution, nor the matrix make-up. Macropores consequent upon the activity of worms, plant roots and other biota have rendered the finer-textured Manawatu soil very permeable at saturation. Inspection of $K(\psi)$ suggests that free-water infiltration into this soil would, when at an antecedent potential of just –10 mm, be characterised by a flow-weighted 'mean' pore radius of 2.2 mm. Definitely macroporous! Many orifices approaching this dimension were observed in the field.

Simple inspection of these two $K(\psi)$ curves allows inferences about the relative roles of macropores and the matrix during the more-common field boundary condition of an imposed flux, V_o. The two soils here both have a K_s of about 7×10^{-5} m/s, or 250 mm/hr. It is very unlikely that a flux of this order would ever be encountered, so it is therefore equally unlikely for $\psi_o = 0$ to prevail across the entire soil surface. For the Manawatu soil however, incipient ponding of the matrix will occur whenever V_o exceeds 0.7 mm/hr for some sufficient time. So the matrix will then act as a local source of free-water for macropores. The macropores need not fill completely to accommodate this. The free-water could thus move below the soil surface along the walls of the macropore system. There however, it would most likely be quite-quickly absorbed through the large area of unsaturated subterranean soil 'exposed' by the macropores. Poiseuille-flow analysis of full-macropore transport would overestimate the rapidity and extent of preferential flow of water and solute. Likewise the results from locally-ponded experiments, both in the field or the laboratory, may give undue weight to macropore flow. Especial attention must be given to the surface boundary condition. Whether or not local matrix ponding ($\psi_0 = 0$) occurs is critical.

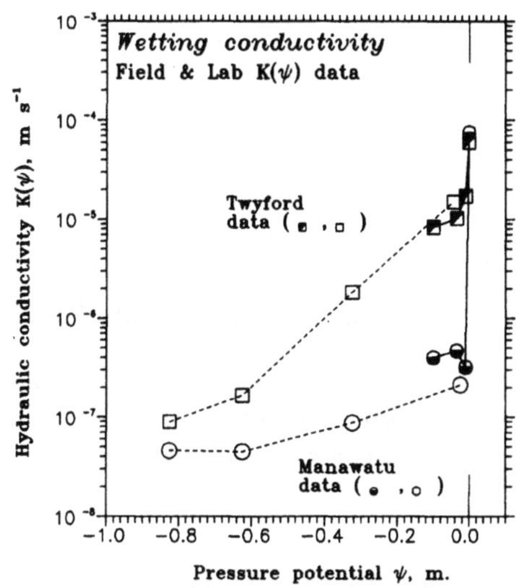

Figure 1. The wetting, unsaturated hydraulic conductivity of Manawatu fine sandy loam (circles) and Twyford sandy loam (squares) as measured by disc permeameters (half-open symbols), and that found by inflow/outflow analysis on undisturbed cores (open symbols).

3. Sprinkler Irrigation.

We studied further the Twyford sandy loam, the more well-behaved soil. Our experiments dealt with sprinkler irrigation, akin to constant-flux infiltration, the least-complicated of boundary conditions. Not surprisingly we just offer a simple field view of this infiltration process.

The laboratory and field $K(\psi)$ data for the Twyford sandy loam can be recast in $K(\theta)$ form (Fig. 2). The data can now be separated into those of the surface 200 mm, and those for the soil below. Whereas both horizons have similarly-shaped $\psi(\theta)$'s, the greater organic-matter content of the surface 200 mm confers on that horizon a higher water content at the same potential.

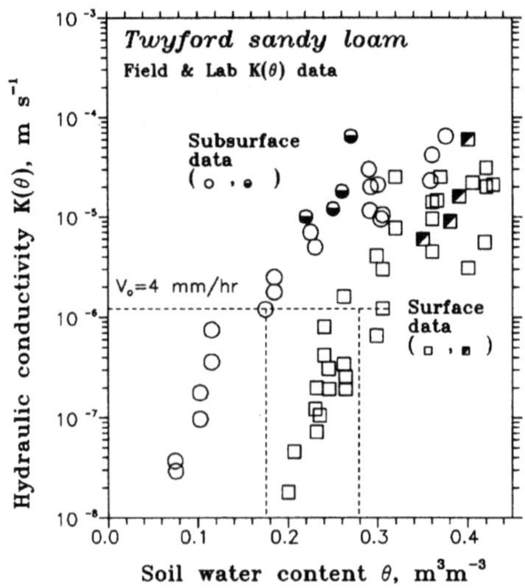

Figure 2. The conductivity functions K(θ) for the surface 200 mm of Twyford sandy loam, and the subsoil. Both data from undisturbed cores (O, ❑) *and disc permeameters* (●, ■) *are shown. The broken lines represents the equilibrium θ^*'s that would result from V_0 = 4 mm/hr.*

On the night of 15th Dec. 1988, the sprinklers in the apple orchard were operated for 10 hours and some 40 mm of irrigation was applied. At a steady, time-averaged rate V_0, it would be expected that behind the wet-front $\partial\psi/\partial z \rightarrow 0$, *viz.* $V_0=K(\theta)$. So the soil water content would asymptote $\theta(z)\rightarrow\theta^*(z)$, where

$$\theta^* = K^{-1}(V_0) \qquad (1)$$

is found from the functional inverse of the hydraulic conductivity relationship (Fig. 2). For the surface soil at this V_0, θ^* = 0.28, and 0.175 for the soil underneath (Fig. 2). The irrigation water infiltrating this soil, initially at θ_n, could simply be expected to fill, at first, the surface 'reservoir' without saturation being achieved. Only then might it proceed to wet the soil underneath. Rational representation of the infiltration process in this manner forms the basis of the oft-used, layered-storage water balance models. Such schemes are sufficiently simple to be usable. Here we consider an unsaturated form. The layer's storage is deduced from the wetting K(θ) and the imposed V_0.

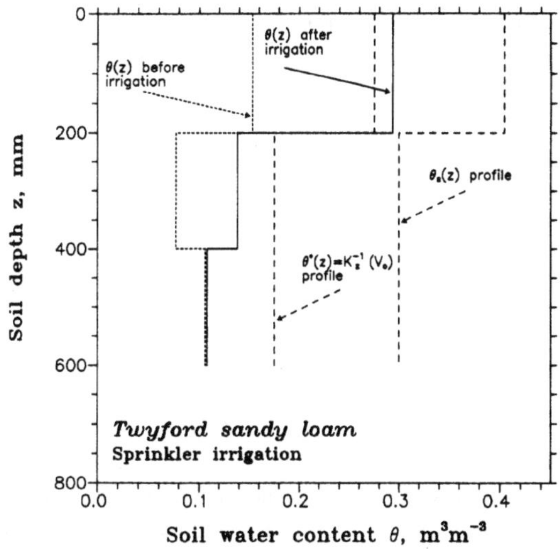

Figure 3. The mean profile of surface soil water content before, and immediately after sprinkler irrigation of an apple tree. The data were obtained by TDR from four sites around the tree. The equilibrium $\theta^*(z)$ and $\theta_S(z)$ come from Fig. (2).

Measurement by TDR at four locations in the bare soil surrounding an apple tree revealed a total irrigation input of 40.3±7.1 mm. This water was entirely recovered in just the top 400 mm of soil (Fig. 3). The variation between sites was not great, as might be expected for an unsaturated matrix, there being no possibility of confounding macropore flow. The surface soil indeed wet to a $\theta^* \approx 0.3 << \theta_S$, as hoped for. The 'excess' proceeded to infiltrate deeper. Yet the wet-front did not progress beyond 400 mm, there being sufficient 'capacity' in the soil between 200 and 400 mm ($\theta < \theta^*$). Such a coarse representation of infiltration seems reasonable, at least for practical purposes. Field measurement of soil wetting during infiltration at a finer scale than that observed here by TDR, would need a great deal of work. Verification, and general application of complex models of field soil-water flow will require an inordinate amount of effort.

4. Root Uptake Flexibility.

These surface observations of soil water content were complemented by 4 neutron-probe $\theta(z)$ measurements down to z^*=1.8 m. Measurements were made about every other day for four weeks following this irrigation. Another wetting event of irrigation, plus some rainfall immediately after(!), occured between the evening of 29th Dec. and 1st Jan. 1989. These data are now used to reveal the spatial pattern of root water extraction.

If the soil water content around a tree, $\theta[x,y,z]$, is monitored down to z^* at several locations $[x,y]$, the spatial pattern of the profile of soil water storage, in mm, is

$$S[x,y] = \int_0^{z^*} \theta[x,y,z]\, dz, \tag{2}$$

so that differentation in time, and integration in space gives the total rate of tree water-use, in litres/day,

$$q = \iint_A \frac{\partial S[x,y]}{\partial t}\, dA\,. \tag{3}$$

Here A is the unit area occupied by each tree. It is assumed that there is neither evaporation from the soil surface, nor drainage below z^*.

The 17 year-old apple trees in this orchard were planted 3.6 m apart, in rows separated by 5 m. Our observations of $\partial S/\partial t$ down to $z^* = 1.8$ m, possessed no spatial structure, rendering trivial the integration of Equation 3. This was not the case when we studied young kiwifruit with an areally-incomplete root system (see Clothier, 1989). There we detected a strong radial pattern in $\partial S[r]/\partial t$ with distance r away from the vine. But here simple averaging of our $\partial\theta[x,y,z,t]/\partial t$ observations yielded the mean rate of apple tree water-use over two periods (Table 1). These rates are in reasonable agreement with the average flux directly determined from summation of half-hourly measurements of sap-flow made within the trunks of two trees by the heat-pulse technique (Green and Clothier, 1988) (Table 1). It is not surprising that evaporation and drainage were negligible. The ground surface was 50% bare soil and to a large extent shaded by the trees. At depth, just beyond 1.8 m, the soil becomes coarser. Unsaturated drainage beyond will become impeded.

Thus instilled with faith that our soil-water content observations indeed represent root uptake, we can analyse the depth-wise pattern of extraction. In Fig. (4) the rate of soil water change q[z] is given in litres/day per tree for slabs of $\Delta z = 200$ mm, over 12 days following the irrigation of 40 mm, and some 8 days after a soil wetting of 67±5.6 mm. Also presented there is the mean profile of root length density deduced from 35, 1.8m long cores exhumed around two neighbouring trees (K.A. Hughes, *pers. comm.*). The two patterns of root uptake are quite different, and neither is in direct proportion to the root length density. Over the twelve days following wetting of just the top 400 mm by the first irrigation of 40.3±7.1 mm, the apple tree withdrew 31.4±3 mm from that surface zone. In order to obtain the complementary 12.8±2.4 mm the tree sought deeper water. Nonetheless some 70% of the extraction occured in top metre, in far greater proportion to the root length density. The second wetting of 67.2 mm resulted in more near-surface water being available. The tree altered its extraction strategy to one of even greater surface preference.Now 87% was taken from the top metre. Gravity and capillarity have allowed some water

Table 1: The rate of apple tree water use q [l/day], over two consecutive periods following soil wetting. Measurements from evaluation of the soil water balance Eq. 3, are compared with those obtained by the heat-pulse sap flow technique.

	Soil water balance $q = \iint_A \left(\int_0^{1.8} \frac{\Delta\theta[x,y,z]}{\Delta t} dz \right) dA$	Heat-pulse sap-flow Tree 6	Tree 10
16-28 Dec 88	80.3 ± 12.9	64.6 ± 3.2	70.8 ± 3.5
2-10 Jan 89	64.3 ± 11.2	57.6 ± 2.9	66.6 ± 3.3

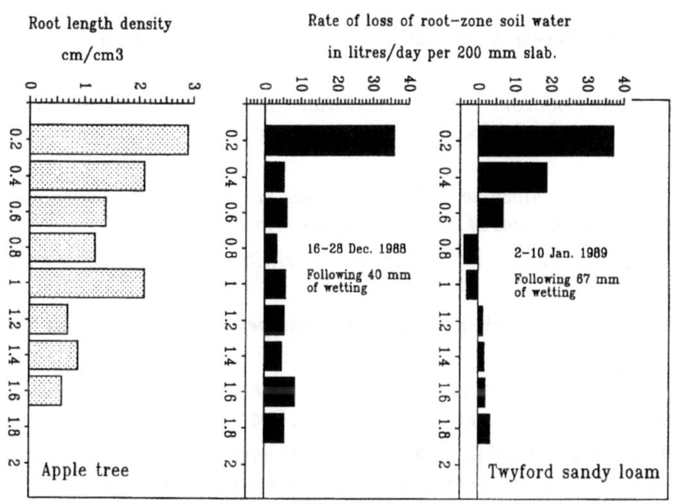

Figure 4. The vertical pattern of apple tree root length density and the rate of soil water extraction per tree, observed following two different soil wetting events.

to escape the preferential near-surface uptake of water from the top 600 mm. In this case the downward redistribution would most likely be contained within the root zone. However in 1987 we observed that a heavy wetting of 109.3 mm resulted in the passage of some 55.7 mm into the coarser underlay beyond 1.8 m (Clothier, 1990). These trees thus appear to prefer to extract first from surface horizons. Any 'excess' water that has been applied then seems to redistribute downwards from the surface, past apparently quiescent roots. It is thus consigned to groundwater, along with any nutrients and herbicides it collected *en route*. Root uptake appears more dependent upon the surface availability of sufficient water, than related to the profile of root length density. Below 1 m during both periods, the soil water content ($\theta \approx 0.15 \pm 0.03$) and the root length density were similar. However the local rates of root water uptake were quite different (Fig. 4). At depth during the first period, $\partial\theta/\partial t$ was about 0.0017 m^3/m^3 per day. The week immediately after however, $\partial\theta/\partial t$ was some threefold less at 0.0006 m^3/m^3 per day. This lower rate of deep extraction followed the more-substantial wetting of the surface horizons of the rootzone. Simple depth-wise models of root water extraction in direct proportion to root length density and in inverse relation to soil water content, would here appear destined to failure. We find the amount of water exiting the rootzone is the result of a complex interaction between the soil's physical properties, and the tree's decision as to where to drink.

References

Clothier, B.E., 1989: *Research imperatives for irrigation science*. J. Irrig. Drain. Engin., **115**, 421-448.

Clothier, B.E., 1990: *Root zone processes and water quality: The impact of management*. Proc. Intern. Symp. Water Qual. Modeling of Agric. Non-point Sources, Logan, Utah 19-23, 1988, USDA, 659-685..

Clothier, B.E. and K.R.J. Smettem. 1990: *Combining laboratory and field measurements to define the hydraulic properties of soil*. Soil Sci. Soc. Am. J. [in press].

Green, S.R. and B.E. Clothier, 1988: *Water use of kiwifruit vines and apple trees by the heat-pulse technique*. J. Exp. Bot., **39**, 115-123.

Perroux, K.M. and I. White, 1988: *Designs for disc permeameters*. Soil Sci. Soc. Am. J. **52**, 1205-1215.

Scotter, D.R. and B.E. Clothier, 1983: *A transient method for measuring soil water diffusivity and unsaturated hydraulic conductivity*. Soil Sci. Soc. Am. J. **47**, 1068-1072.

Smettem, K.R.J. and B.E. Clothier, 1989: *Measuring unsaturated sorptivity and hydraulic conductivity using multiple disc permeameters*. Soil Science, **40**, 563-568.

Brent Clothier, Environmental Physics Group, D.S.I.R.: F & T, Private Bag, Palmerston North, New Zealand.

Keith Smettem, Division of Soils, C.S.I.R.O., Private Bag 2, Glen Osmond, SA 5064, Australia.

Pudjo Rahardjo, Research Institute for Tea and Cinchona, P.O. Box 148, Bandung, Indonesia.

Field-Scale Water and
Solute Flux in Soils
Monte Verità

THE INFILTRATION-OUTFLOW EXPERIMENT USED TO DETECT FLOW DEVIATIONS

M. Císlerová, T. Vogel and J. Simunek

Simple recurrent infiltration-outflow experiments, together with dye tracing, show deviations from the theoretical assumptions on which classical soil-water flow modelling is based. The experimental set-up represents a well-defined system which enables the study of the dynamic character of soil properties expressed in the Richard's-equation approach by soil hydraulic functions. The method may offer the possibility to develop an efficient classification of soils according to their dynamic behaviour. Results from four soils are presented.

1. Introduction

Significant deviations from the presumed behaviour of water flow in soil can take place, but often they are hidden by soil variability effects. It is important to know more about the character of these deviations, since many of them contradict the theoretical assumptions on which soil-water modelling is based. It is generally suggested that these phenomena should be looked at in the field. However, to perform field experiments in sufficient detail, to have control over an area of flow with well-defined boundary conditions, requires huge investments and several months of effort. For the detection of deviations, simple infiltration-outflow experiments on large undisturbed core samples, followed by dyeing of the soil core, can supply substantial information. From the measured results, the character of flow can be deduced. We suggest to use this type of experiments for a classification of soils according to their flow. Of special importance we see is the chance to detect the preferential flow. We have the feeling that such a kind of classification is badly needed. It is necessary to find a common language when speaking about such things as model-validation, variability effects, inverse problem solution, etc.

2. Experiment description

The experiments here were done on cylinders of 200 mm of diameter and 200 mm long (Císlerová et al., 1988). On each undisturbed soil sample ponding infiltration experiments were performed in the laboratory. Pressure heads were checked either by three tensiometers installed at various heights, or simply by one in the centre of the soil sample. The outflow was measured at the bottom of each sample. Water was supplied in small regular doses to keep the ponded water level at approx. 5 mm in height. To avoid disturbance, the surface was covered by a small sheet of cloth. At the bottom a fine mesh hold by a frame supported the soil core. The outflow volumes were collected in a fraction collector at regular intervals. After the first infiltration experiment into the relatively dry soil, several repetitions were performed, with lags between 20 minutes and one day.

With respect to the flow, such a set up represents the well-defined system with the constant head boundary condition at the top, and a free water boundary condition at the bottom. The initial condition water content was determined from the changes of the weight of each sample, and the tensiometers.

To obtain information about the effective soil area which took part in the flow, finally a non-reactive blue dye was infiltrated for sufficient time to achieve a fully-coloured outflow. Subsequently, when the soil was dry enough to be manipulated, usually after few days, the sample was sliced in order to measure the dye distribution.

3. Example soils

Results of infiltration-outflow experiments are presented for 4 different soils: They are:

Korkusova Hut	(Bohemia)	- sandy loam,
Hupselse Beek	(Holland)	- fine sand,
Braunschweig	(W.Germany)	- clay loam,
Trebon	(Bohemia)	- gravelly sand.

Each of the soil represents an experimental area where detailed field observations have been done. Korkusova Hut is a part of the experimental watershed of the Czech Hydrometeorological Institute situated in the National Park Sumava (Císlerová et al., 1988). Hupselse Beek experimental watershed belongs to the Agricultural University in Wageningen (Hopmans and Stricker, 1987), Braunschweig represents the experimental area of the Agricultural University in Braunschweig (Othmer and Bork, 1989). In the Trebon region a large project is under way to detect vulnerable areas with the respect to groundwater protection from surface source pollution. All samples were taken from the second horizon. In all four cases no visible preferential ways like wormholes or cracks could be seen. The particle size distribution of particular soils is evident from Fig. 1. The examples of the infiltration - outflow results are shown in Figs. 2 to 5. These can be compared with the corresponding dye-distributions given in Fig. 6.

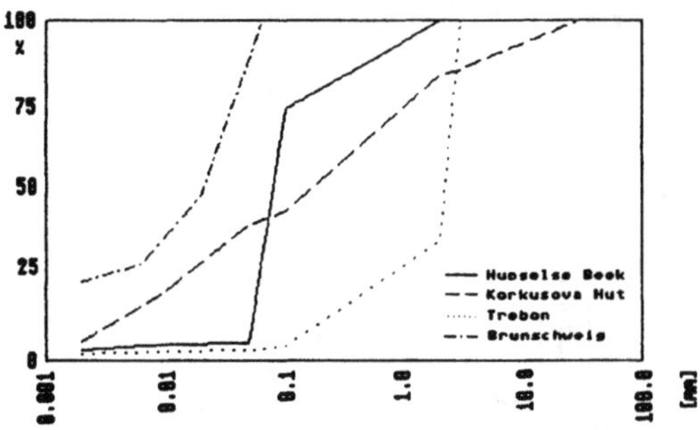

Figure 1: Particle-size distributions.

4. Discussion of Results

4.1 Korkusova Hut

In Fig. 2, the infiltration run with one repetition is presented. During the first run, water was infiltrated into the relatively dry soil. For the second, the initial moisture content distribution would have been the result of redistribution following the first. In both cases the outflow started within a very short time of infiltration, and the outflow rate rapidly became equal to the infiltration rate. Both were constant in time. This steady-state flux represents the saturated hydraulic conductivity. Consider the ratio of infiltration, or outflow rates of two runs. The ratio between the rate of second run to the previous run was just 0.5. This is also true for the volume of water flowing out of the sample in post-infiltration stage. The volume was always counted from the time at which the ponded water layer on the surface of the sample vanished. This observation that the post-infiltration cumulative outflow shows a significant decrease from the measured volume in the second run is not in agreement with established, uniform theory. This particular case has been documented in more detail in the paper of Císlerová et al. (1988), together with field experiments where the same effect has been shown. From the dye-distribution results given in Fig. 6 it is evident that only a part of the soil volume took part in the flow. Here the more black the colour the more dye in the soil. The rest of the soil volume was nevertheless wet, but without any traces of colour. There were no visible disturbances or continuous pathways.

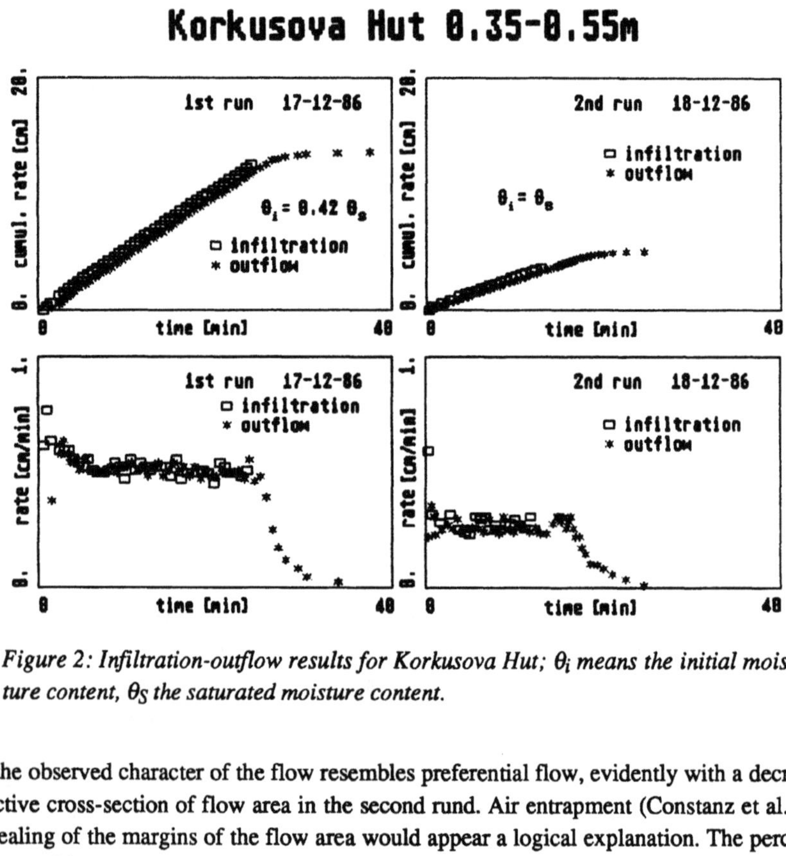

Figure 2: Infiltration-outflow results for Korkusova Hut; θ_i means the initial moisture content, θ_S the saturated moisture content.

Still the observed character of the flow resembles preferential flow, evidently with a decrease in the active cross-section of flow area in the second rund. Air entrapment (Constanz et al., 1988) and sealing of the margins of the flow area would appear a logical explanation. The percolation theory would seem to be convenient in this case (Wardlaw, 1988, Luxmoore et al., 1989).

4.2 Hupselse Beek

As can be seen in Fig. 3, during the first run with the dry sample, the outflow started 23 minutes after the beginning of infiltration. Soon after, both the infiltration and outflow rates became equal. They increased slightly as air bubbled out under the pond. The repetition into the nearly saturated sample resulted in an unchanged infiltration rate and rate of outflow, the latter of which however started immediately. There was no post-infiltration outflow at all. The picture of dye-distribution shows the very uniform distribution of dye all over, except below tiny, local fissures. These fissures were not visible, but above them, the accumulation of organic matter had created spots of darker colour. Still the character of the flow in this case seems to correspond to the classical theory of water movement through the pore space.

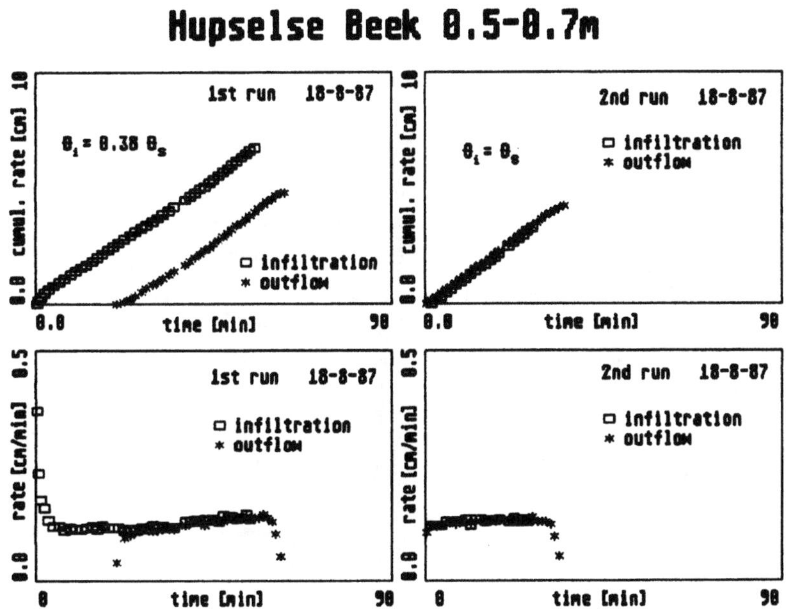

Figure 3: Infiltration-outflow results for Hupselse Beek; θ_i means the initial moisture content, θ_S the saturated moisture content.

4.3 Trebon

Another type of flow is demonstrated in Fig. 4. Outflow started immediately. From the equal rates of infiltration and outflow it can be concluded that preferential flow was dominant. Both rates decrease constantly. In another un-shown repetition the steady-state infiltration rate only reached 1/3 that of the first run. The post-infiltration flow was not recorded. The dye-distribution shows that the flow was significantly preferential and occurred through complicated random three-dimensional network of coarse pores (see Fig. 6) where again no signs of structural macro-pores could be seen. The decrease in the flux during the infiltration was possibly due to swelling of clay particles.

4.4 Braunschweig

Firstly, note that the time scale of this experiment is very different from the previous cases. The outflow started more than one hour after commencement of infiltration into this relatively dry soil (Fig. 5). Afterwards the outflow rate reached the infiltration rate. The simultaneous and gradual decrease in both the infiltration and outflow rates continued. Post-infiltration outflow

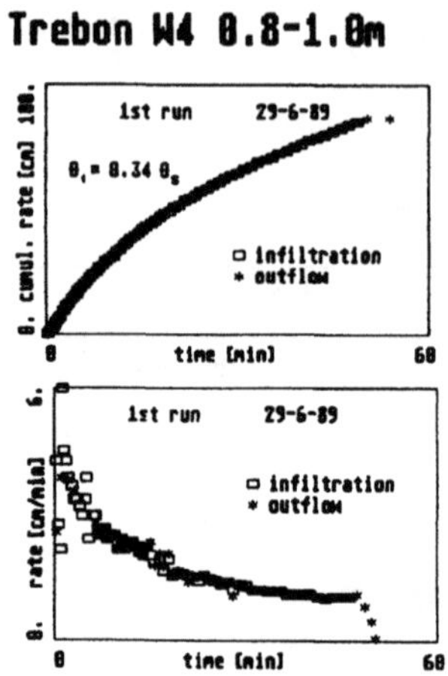

Figure 4: Infiltration-outflow results for Trebon.

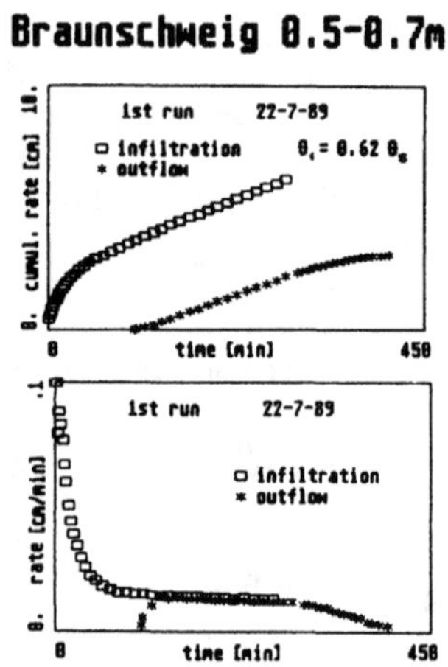

Figure 5: Infiltration-outflow results for Braunschweig.

(θ_i means the initial moisture content, θ_S the saturated moisture content)

lasted for almost two hours. During infiltration the soil was evidently swelling, at the end of experiment it had a jelly-like consistency. No further runs could be performed. In detailed characterization of this soil (Othmer and Bork, 1989), both the subangular blocky structure and frequent macropores (wormholes) are mentioned. Here, however, the flow apparently represented regular matrix flow through a swelling medium.

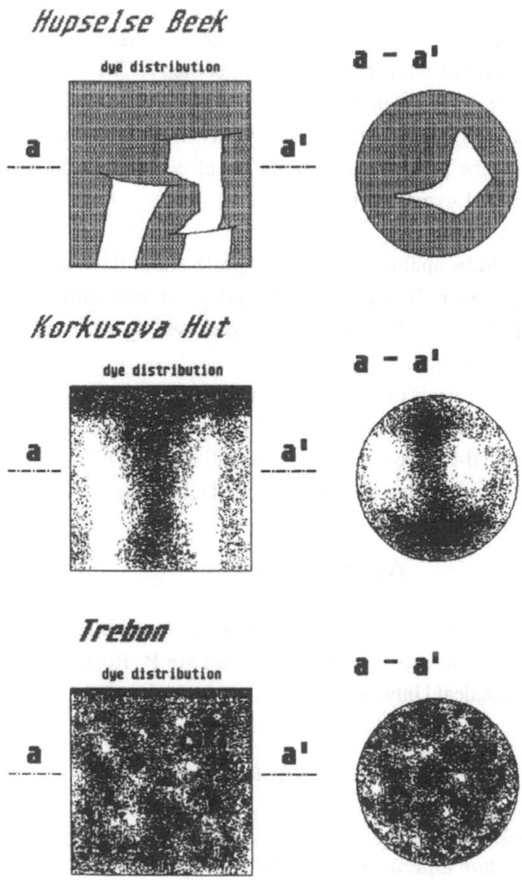

Figure 6: Dye-distribution images; vertical and horizontal transects respectively

5. Conclusions

The results here confirm that for certain soils, classical determination of soil's hydraulic characteristics is insufficient. In the examples presented, the dynamic and variable nature of soil hydraulic properties can be seen. A dependence of the flux of water on the wetting history was observed. In the case of ponding infiltration this represents a non-uniqueness in the saturated hydraulic conductivity. It is evident that these effects are characteristic features of each soil, so it means they should be classified with respect to soil geometry. The preferential flow was detected. Dye tracing provides an image of the flow area of the preferential pathways. In two cases the detected instability of saturated hydraulic conductivity in combination with the irregularities of the flow area witnesses the presence of percolation effects in dependence on the pore size distribution (Wardlaw, 1988, Luxmoore et al., 1989).

The method used here could be applied to detect the preferential flow locally, and help to classify the modes of its appearance. It could help to specify the preferential flow due to soil features, in contrast to the preferential flow that might be caused by secondary influences such as climate, plants, animals, human activity.
We suggest that simple experiments of this type, as well as ones performed using modern laboratory technique, perhaps with triple dye-images and computer-assisted analysis (Crestana et al. 1988), might bring the looked-for progress in understanding of the field soil-water system. They reveal the behaviour of flow in dynamic but well-defined situations.

Acknowledgements

We are grateful for cooperation to J.N.M. Stricker from LU Wageningen and to H. Othmer from LU Braunschweig. The other members of working team are R. Röslerová, J. Janatková and J. Valentová from Czech Technical University, Praha.

References

Císlerová, M:, J. Simunek and T. Vogel, 1988: *Changes of steady-state infiltration rates in recurrent ponding infiltration experiments*. J. of Hydrology, **104**, 1-16.

Constantz, J., W.N. Herkelrath and F. Murmphy, 1988: *Air encapsulation during infiltration*. Soil Sci. Soc. Am. J., **52**, 10-52.

Crestana, S., S. Mascarenhas, R. Cesareo and P.E. Cruvinel, 1988: *Soil research opportunities using X- and gamma-ray computed tomography techniques*. Proc. Validation of Flow and Transport Models for the Unsaturated Zone, co-editors P. Wierenga and D. Bachelet, Ruidoso.

Luxmoore, R.J., G.V. Wilson, P.M. Jardine and R.H. Gardner, 1989: *Use of percolation theory and Latin hypercube sampling in field-scale solute transport investigation.* This proceedings.

Othmer, H., H.R. Bork, 1989: *Characterization of the soils at the investigation sites.* Landschaftsgenese und Landschaftsökologie, Braunschweig, **16**, 73-86.

Wardlaw, N.C., 1988: *Fluid topology, pore size and aspect ratio during inhibition.* Transp. in Por. Media, **3**, 17-34.

Milena Císlerová, Dept. of Drainage and Infiltration, Faculty of Civil Engineering, Czech Technical University, Thakurova 7, 166 29 Praha 6, Czechoslovakia.

T. Vogel and J. Simunek, Research Institute for Improvement of Agricultural Soils, Bozantní 697, 16500 Praha 6, Czechoslovakia

Field-Scale Water and
Solute Flux in Soils
Monte Verità

SPATIAL VARIABILITY OF UNSATURATED FLOW PARAMETERS IN FLUVIAL GRAVEL DEPOSITS

F. Stauffer and P. Jussel

Unweathered outcrops of fluvial gravel deposits near Zürich have been investigated in order to obtain information about inhomogeneities of hydraulic parameters. Different geological elements are distinguished and analysed on photographs. The variability of unsaturated flow parameters is studied based on measurements of the porosity and the grain size distribution using the program SOILPROP. A special method is applied in order to estimate the desaturation characteristics of gravel samples. The method is tested with data for undisturbed soil samples. Each geological element shows a stochastic variability which strongly contrasts with that of other elements. The purpose of the study is a geostatistical description of the unsaturated flow parameters. The effects of spatial variability on predominantly vertical unsaturated flow and transport processes are discussed. Unsaturated flow and transport in the gravel deposit is expected to be dominated by the geometrical arrangement of the open framework zones and sand lenses.

1. Introduction

The gravel aquifers of the prealpine valleys in Switzerland represent an important drinking water reservoir due to their volume and their high values of the hydraulic conductivity. Since the natural replenishment of these aquifers is obtained by infiltration of river water and the areal recharge of rainwater, the vadose zone above the groundwater table forms an important buffer zone for water quality reasons. The transport processes of soluble substances within the vadose zone are influenced by the hydraulic characteristics of the unsaturated domain (water content-capillary pressure-hydraulic conductivity relationship), molecular diffusion and mechanical dispersion as well as chemical sorption and reactions. An important feature of the gravel deposits is their high degree of non-homogeneity in the textural composition. The non-homogeneity in the hydraulic characteristics largely influences the flow as well as the transport behaviour.

The aim of this study is an investigation of the spatial variability of the unsaturated hydraulic characteristics within gravel deposits. The study involves a classification of different geological elements which are discussed from the hydraulic point of view. The calibration of the different geological elements is undertaken by investigating a series of disturbed and undisturbed soil samples. Parameters of the unsaturated hydraulic characteristics are estimated based on porosity and saturated hydraulic conductivity measurement of the undisturbed samples, and their grain size distribution. The study should further give an answer to the question whether the consideration of the geological elements is relevant from a hydraulic point of view. The statistical description finally will represent a basis for a model formulation of flow and transport processes in gravel deposits.

2. Characterization of gravel deposits

The investigated fluvial gravel deposits are situated in front of a Würmian moraine which has not been deformed by a glacier during the last glacial period. The data was collected in the Hüntwangen gravel pit situated in the north of Zürich (Jussel, 1989). The thickness of the unsaturated zone can be as large as 40 m. Based on a geological description of the outcrops (Huggenberger et al., 1988), the deposits may be classified into six characteristic types of lithological and textural units denoted here as geological elements:

1. Grey gravel: Sandy, poorly sorted gravel with little silt, marginal or no layering, grey (blue) colour.
2. Brown gravel: Sandy and silty, poorly sorted gravel containing pebbles, some layering. The silt causes a brown colour.
3. Sand lenses: Well-sorted sand which is present as lenses, little silt, grey colour, rather strong homogeneity within the lense.
4. Open framework gravel: Well-sorted gravel with practically no sand or silt; occurs usually in very thin horizontal or inclined lenses.
5. Open framework / bimodal couplets: More or less regular alternations of inclined open framework zones with bimodal gravels. The bimodal gravels usually consist of coarse gravel or pebbles in a matrix of well-sorted medium sand.
6. Silt lenses: Well-sorted silt, occurs in lenses. Brown colour. Rarer than sand lenses.

Typical grain size distribution curves are shown in figure 1. Almost all of the gravel grain size distribution curves show a more or less distinct bimodal behaviour. Grey and brown gravel can be considered to be the basic structure, whereas the rest of the elements are lense-shaped inclusions in this base matrix. Mixtures of the described geological elements also exist. The classification was confirmed in a series of sites of gravel deposits in Northern Switzerland. The spatial distribution of the different elements was investigated with the help of photographs (Jussel, 1989). A sketch of the geological deposits of a typical outcrop in the Hüntwangen pit is shown in figure 2. The areal or volumetric fraction of the different elements is given in table I.

The assumption of a perfect or almost perfect layering does not hold for the investigated gravel deposits. The typical structure consists of lenses of various hydraulic conductivities, inclinations and extents. The hydraulic behaviour for the elements under saturated conditions was determined by the laboratory measurement of a total number of 35 small undisturbed samples (gravel samples: cube of 20 cm; sand samples: diameter of 10 cm, length 20 cm). Mean and standard deviation of the measured porosities and hydraulic conductivities (logK, K in m/s) are listed in table II. The corresponding normal distributions are shown in figure 3 and 4. The results indicate that the parameters of the different elements can clearly be distinguished. The optical classification thus reflects the differences in the hydraulic behaviour.

Table I: Volume fraction of the different geological elements.

geological element	% of outcrop
grey gravel	11.2
brown gravel	24.3
Mixture of grey and brown gravel	44.9
Mixture of grey gravel and sand	3.1
sand lenses	6.9
open framework zones	1.5
openwork/bimodal-couplets	7.7
silt lenses	0.4

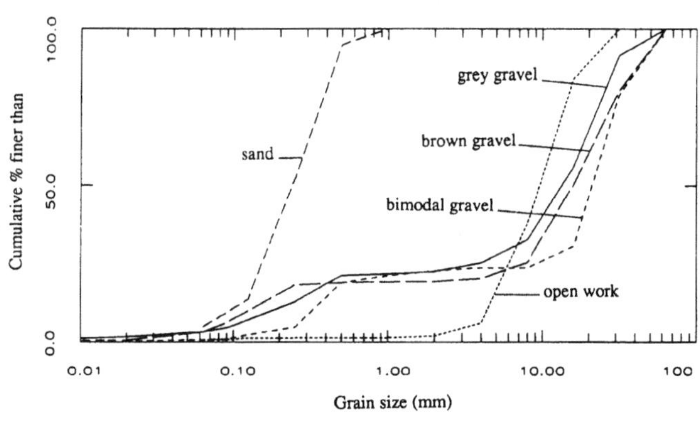

Figure 1: Typical grain size distribution curves.

The saturated hydraulic conductivity in both the horizontal and vertical directions was determined for undisturbed grey and brown gravel samples (Jussel, 1989). The set-up used corresponds to a constant head permeameter. The results show that practically no or only a small degree of anisotropy exists within a scale of 20 cm. The relation of horizontal to vertical hydraulic conductivity was: $0.6 < K_{hor}/K_{vert} < 1.5$.

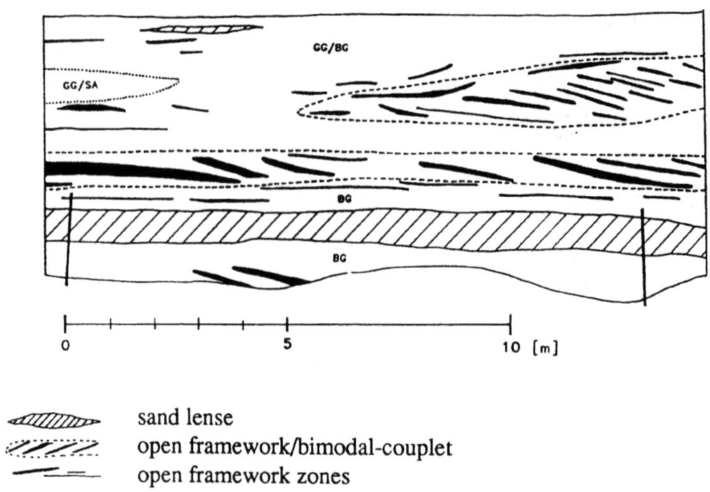

Figure 2: Sketch of the geological deposits of a typical outcrop in the Hüntwangen pit; BG: brown gravel; GG/BG: alternating thin layers of grey and brown gravel; GG/SA sandy grey gravel.

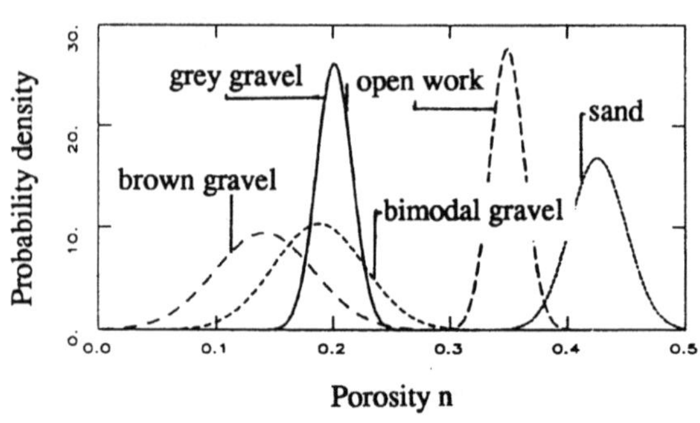

Figure 3: Probability density function of measured porosity n.

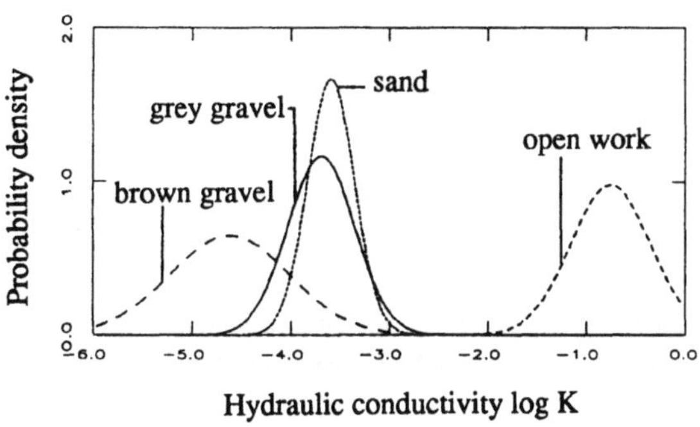

Figure 4: Probability density function of measured saturated hydraulic conductivity K.

The correlation structure of the hydraulic conductivity (lnK) was investigated for a wall of grey gravel only (Jussel, 1989). For this purpose 40 disturbed samples were taken in intervals of 1.2 m along a horizontal line. Saturated hydraulic conductivity was determined based on the grain size distribution and the method of Kozeny (Langguth and Voigt, 1980). The method itself was calibrated by measurements on the undisturbed samples. The resulting estimation for the correlation length of lnK was 2.8 m and the standard deviation σ_{lnK} was 0.14.

3. Estimation of unsaturated flow parameters

The aim of the analysis is a parametrization of the desaturation characteristics. Estimates of the capillary pressure - water content relationship may be obtained by the method of Arya and Paris (1981) using measurements of the grain size distribution and of the porosity. The method assumes that any grain size fraction is associated with a portion of void space which is proportional to its weight. By a conceptual model the capillary pressure within that void space is estimated based on the grain size diameter of the fraction. The approach was successfully applied by Mishra et al. (1989) to a number of soil samples. However, the applicability of the method on gravel samples has to be critically reviewed. The usual procedure is to subtract the stone and pebble fraction and to restrict the approach to the remaining grain size classes. The approach seems to be reasonable since very large pores originating from pebbles and stones are often completely filled with finer material and therefore do not affect the desaturation behaviour with the exception of a modified porosity. For a given stony fraction the porosity of the remaining fine material is:

$$n' = \frac{n}{1 - p \cdot (1-n)} \qquad (1)$$

where n is the porosity of the original sample, n' the porosity of the fine material, and p the ratio of the weight of pebbles and stones to the total weight of the dry sample. A crucial question is the indication of the maximum grain diameter d_p associated with p. The adoption of a maximum diameter corresponding to usual grain size classification schemes seems to be arbitrary in this context.

An estimate for d_p may be obtained by assuming that the pore volume of the coarse material would be completely filled with the fine material. Assuming that the porosity of the coarse material equals that of the fine one, which imposes geometrical similarity of the fine and the coarse materials, it can be shown that $n' = \sqrt{n}$. However, the latter assumption can rarely be verified, so that this approach remains uncertain. Nevertheless a measure of p may be obtained. The resulting value for d_p is around 8 mm for the gravel samples with the exception of the open framework gravel.

A different concept for the determination of d_p would take advantage of the bimodal character of the grain size distribution curve of the gravel samples. For most of the samples a distinct separation between the fine and the coarse grain diameters occurs around 0.5 mm. However, the resulting porosity n′ is too high (0.6 or higher).

As a compromise the maximum grain size diameter d_p is chosen so that the porosity of the fine material n' lies in a reasonable interval taking into account the non-uniformity of the grain size distribution ($n' \cong 0.3$-0.5). The resulting diameter d_p is 8 mm for most of the samples. Based on d_p the estimation of the desaturation curve was performed with the program SOILPROP (Mishra et al., 1989).

Although the method could not be tested directly with the undisturbed samples mentioned above due to limited means of the study, it was tested with measured data on undisturbed gravel samples of similar geological sites of Richard et al. (1983, sample Winzlerboden) and Vogelsanger (1983, samples K4 and K5). The estimated desaturation curve is in reasonable agreement with the data at smaller and medium water contents for all samples. The predicted capillary pressures are strongly underestimated for the higher saturations. The reason is the wide range of particle sizes and the fact that the pores are interconnected. Trapped large pores are drained only at significantly higher capillary pressures. The deviation from the predicted curve is characterized by a distinct decrease of the capillary pressure which causes a pronounced edge in the curve. The related saturation at the edge was around 0.5. This behaviour was observed for all investigated gravel samples. The predicted desaturation curve is therefore accepted only up to a saturation lower than that at the edge. This approach is further supported by the fact that the water content at the edge corresponds to a grain diameter class which is partially lacking. The unsaturated characteristics are thus mainly determined by the fine material. Based on a number of estimates of the capillary pressure and water content by means of SOILPROP the parameters h_b, λ and S_r of Brooks and Corey relation (Corey, 1977):

$$\frac{S - S_r}{S_m - S_r} = \left(\frac{h_b}{h_c}\right)^{\lambda} \tag{2}$$

are determined, where S_m is the maximum and S_r is the residual saturation, h_c the capillary pressure head, and h_b the air entry pressure head. S_m is chosen as 0.911 according to Mishra et al. (1989). The parameters are determined by non-linear optimization (Stauffer and Dracos, 1986). The predicted S_r was practically zero for all soil samples. The estimated desaturation function is shown together with the measurements for the test samples in figure 5. For higher saturations than around 0.5, the curves represent extrapolations.
The data show the quality of the present estimates of the desaturation function. In test sample K4 the residual saturation S_r is strongly underestimated. Nevertheless it is assumed that the parameters h_b and λ represent the unsaturated flow behaviour of the samples.

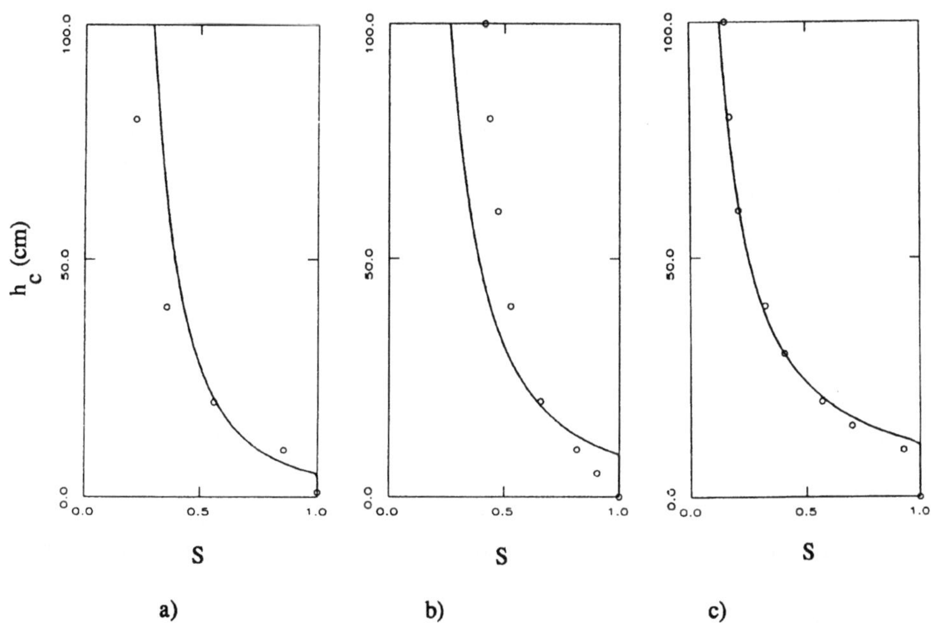

Figure 5: Relationship between capillary pressure head h_c and saturation degree S for gravel samples; a) Winzlerboden, b) K4, c) K5; solid line: predicted Brooks-Corey curve; points: measured data of Richard et al. (a, 1983) and Vogelsanger (b,c, 1983).

4. Stochastic description of unsaturated flow parameters

Based on the data of 35 undisturbed gravel and sand samples from the Hüntwangen pit the estimates of the retention parameters are determined by the method described above. The probability densities of h_b and λ are shown in figure 6 and 7. The data for the open framework samples only allowed a crude estimate of the desaturation parameters (h_b less than 1 mm and $\lambda \cong 1$). The values were therefore disregarded in the figures. The parameters of the lognormal probability density (mean and standard deviation) are listed in table II.

5. Discussion

The proposed method for estimating the desaturation parameters is applicable for the grey and brown gravel as well as the sand samples. Only crude estimates are obtained for the very coarse open framework samples. The parametrization of the hydraulic characteristics according to Brooks and Corey together with the hydraulic conductivity at saturation and the porosity provide a simplified but complete set of the essential unsaturated flow parameters. They characterize the relations between the capillary pressure, the water content and the hydraulic conductivity.

Table II: Parameters of the probability density of n, logK, h and λ (mean and standard deviation σ).

Material	number of samples	$\overline{n}$	σ_n	$\overline{\log K}$	$\sigma_{\log K}$	$\overline{h_b}$ (cm)	σ_{hb}	λ	σ_λ
grey gravel	6	0.20	0.015	-3.7	0.34	12.0	6.8	0.60	0.28
brown gravel	6	0.14	0.042	-4.6	0.61	18.4	27.3	0.46	0.26
sand	13	0.43	0.024	-3.6	0.24	40.3	10.8	1.35	0.31
bimodal gravel	3	0.19	0.039			19.7	10.8	0.89	0.32
open work gravel	7	0.35	0.014	-075	0.41				

The statistical analysis of the data shows that the different geological elements can clearly be distinguished according to their hydraulic properties. Key parameters like the porosity and/or the saturated hydraulic conductivity are characterized as expected by much less variation within a geological element than between elements. Gravel samples (grey, brown and bimodal gravel) can clearly been distinguished from sand and open work samples (Fig. 2, 3, Table II). The same holds but to a lesser extent for the estimates of the desaturation characteristics (h_b and λ, Fig. 5, 6, Table II).

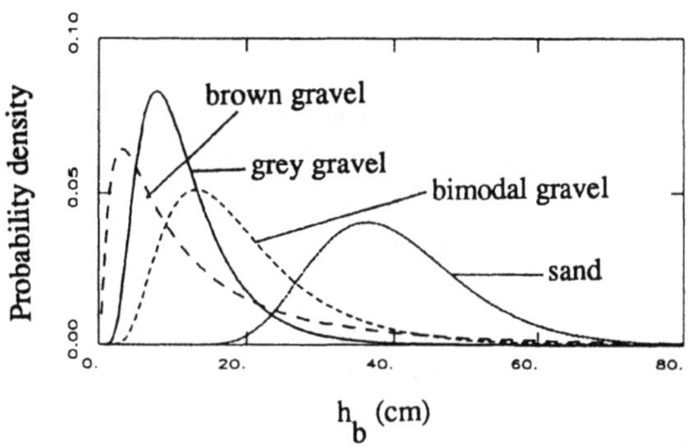

Figure 6: Probability density of estimated h_b of gravel and sand samples (Hüntwangen pit).

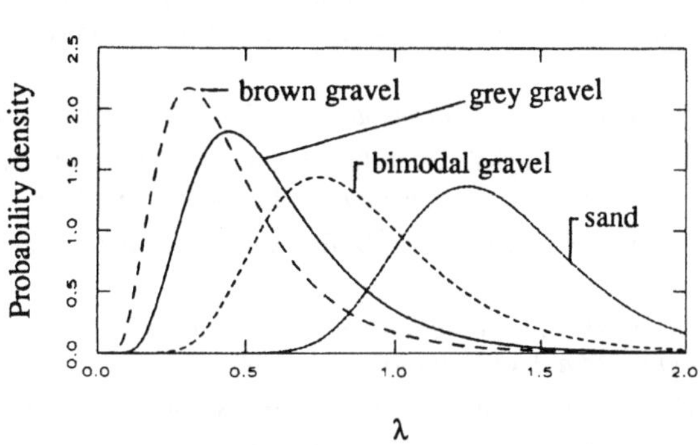

Figure 7: Probability density of estimated λ of gravel and sand samples (Hüntwangen pit).

The correlation length of saturated hydraulic conductivity estimates was determined along a horizontal line of grey gravel. The corresponding vertical correlation scale seems to be reduced according to an observation of the outcrop. The correlation behaviour of the overall outcrop for all parameters is dominated by the structure formed by the different geological elements. This incorporates also the anisotropic behaviour. For a scale larger than that of a sample (scale of 20 cm) anisotropy is evident. It again stems from the overall structure of the geological elements which is dominated by more or less horizontal lenses and layers.

An important feature for the unsaturated flow and transport behaviour is given by the arrangement of the open framework zones. Horizontal layers may act as hydraulic barriers whereas inclined layers together with bimodal couplets may provide horizontal displacement components. The sand lenses with their rather high values of both hydraulic conductivity and air entry pressure also contribute to an increase of the horizontal flow components. Consequently the overall arrangement of all these zones strongly influences the mixing behaviour of soluble substances.

References

Arya, L.M., and J.F. Paris, 1981: *A physicoempirical model to predict the soil moisture characteristic from particle-size distribution and bulk density data.* Soil Sci. Soc. Am. J., **45**, 1023-1030.

Corey, A.T., 1977: *Mechanics of heterogeneous fluids in porous media.* Water Res. Publ., Fort Collins, CO.

Huggenberger, P., C. Siegenthaler, and F. Stauffer, 1988: *Grundwasserströmung in Schottern; Einfluss von Ablagerungsformen auf die Verteilung der Grundwasserfliessgeschwindigkeit.* Wasserwirtschaft, **78**, 202-212.

Jussel, P., 1989: *Stochastic description of typical inhomogeneities of hydraulic conductivity in fluvial gravel deposits.* Proc. Int. Symp. on Contaminant Transport in Groundwater, Stuttgart, April 4-6, 221-228.

Langguth, H.-R., and R. Voigt, 1980: *Hydrogeologische Methoden.* Springer, Berlin.

Mishra, S., J.C. Parker, and N. Singhal, 1989: *Estimation of soil hydraulic properties and their uncertainty from particle size distribution data.* J. Hydrology, **108**, 1-18.

Richard, F., and P. Lüscher, 1983: *Physikalische Eigenschaften von Böden der Schweiz.* EAFV Birmensdorf, **Vol. 3.**

Stauffer, F., and T. Dracos, 1986: *Experimental and numerical study of water and solute infiltration in layered porous media.* J. of Hydrology, **84,** 9-34.

Vogelsanger, W., 1983: *Untersuchung über den Wasserhaushalt eines zweischichtigen Bodenprofiles unter Waldbestockung, dargestellt an einer sandigen Parabraunerde über Schotter.* Diss. ETH Zürich, Nr. 7307.

Fritz Stauffer and Peter Jussel, Institute of Hydromechanics and Water Resources Management, ETH-Hönggerberg, CH-8093 Zürich, Switzerland.

This study was partially supported by the Swiss National Science Foundation.

Field-Scale Water and
Solute Flux in Soils
Monte Verità

QUANTIFICATION OF DETERMINISTIC AND STOCHASTIC VARIABILITY COMPONENTS OF SOLUTE CONCENTRATIONS AT THE GROUNDWATER TABLE IN SANDY SOILS

J. Böttcher and O. Strebel

Solute concentrations at the groundwater table in sandy soils measured along transects (30 m) were statistically analyzed to separate and quantify deterministic and stochastic variability components. Results for nitrate concentration under arable land and for sulfate concentration under coniferous forest are presented. Spatial variability of these solute concentrations at the groundwater table is influenced by deterministic variability components. These components represent the following extrinsic factors: Fertilization practices on arable land and, under coniferous forest, input into the soil via throughfall of canopy-deposited atmospheric constituents.

1. Introduction

For description of nonpoint regional groundwater pollution problems, the knowledge of solute input into the groundwater by leaching from the unsaturated soil is essential. This solute input is soil and landuse specific and was characterized by Strebel and Böttcher (1989) by solute concentrations measured at the groundwater table. The evaluation of those data showed the variability of solute concentrations on a "regional scale" (catchment area [order km^2]). This variability is at least partly due to spatial variability on a "local scale" (single field [order m^2 to ha]). To gain insight into the "local scale" spatial variability of the solute input from sandy soils into groundwater, solute concentrations at the groundwater table under arable land and coniferous forest were measured along transects of 30 m length. It was found, that extrinsic factors (related to landuse) strongly influence the variability patterns of solute concentrations at the groundwater table, and thus represent deterministic components (trends, regular fluctuations) in the observed spatial variability. Fertilization practices on arable land and solute input under coniferous forest via throughfall and removal of canopy-deposited atmospheric constituents, must be considered as important factors.

Commonly, spatially predictable variability components are defined as *deterministic variability* (e.g. Davidoff et al., 1986; Russo and Jury, 1987). This definition is valid for both approaches followed in this paper to quantify the deterministic variability components. In addition, for one approach (space series analysis) a more stringent definition can be applied.
Besides the deterministic variability, a spatially unpredictable random variability component was present in our transect data. In accordance with the definition of Russo and Jury (1987), we call this *stochastic variability*.
However, the ratio of deterministic to stochastic variability components of groundwater solute concentrations is not unique, but depends on the scale or sampling interval, respectively. Scale dependence means that a deterministic component, e.g. a short periodic fluctuation, can be assigned to the stochastic variability if the sampling scale or interval is enlarged, and vice versa. Furthermore, it is to be expected that the sampling date (especially in case of solutes with a pronounced seasonal leaching behavior, such as nitrate under arable land), and the direction of sampling transects (in the case of spatially anisotropic behavior) are important factors, that can influence this ratio.
The objective of our paper is the *separation* of the "local scale" variability of solute concentrations at the groundwater table into deterministic and stochastic components using two different approaches and the *quantification* of these variability components as portions of the total variability (= sum of deterministic and stochastic variability).

2. Field measurements

The investigations were carried out in the "Fuhrberger Feld" area, an extensive, flat area in northwest Germany, about 30 km north of Hannover. The soils are Gleyic Podzols, formed in fine to medium sands, with groundwater tables at depths of 80 to 130 cm below groundsurface. The mean annual precipitation is 680 mm/a, and the mean annual temperature is 8.8°C.
Groundwater samples were taken from the first decimeter below the water table in March 1985 at a time period when high seepage rates and high solute concentrations at the groundwater table could be expected. A simple handauger-slitprobe technique was used for sampling. 61 samples (≈ 200 ml) were taken at 0.5 m intervals along 30 m transects under bare arable land (pre-crop maize) perpendicular to the direction of plowing, and under coniferous forest (45 years old pine stand) between two rows of trees. The groundwater samples were analyzed for Na, K, Ca, Mg, Al, NO_3, SO_4, Cl, pH and electrical conductivity. For details see Böttcher and Strebel (1988b,c).

3. Mathematical and statistical methods

To separate deterministic and stochastic variability components of the space series of groundwater solute concentrations two different approaches are used (Figure 1). In the *first approach* (sections 3.1 - 3.4), methods of time/space series analysis are used to identify the deterministic variability components and to assign them to different factors, mostly extrinsic. That means, in approach (1)only those variability components which are (i) *spatially predictable* and (ii) *causally related with a known factor* are considered as deterministic. The methods used are trend verification and elimination (TE), spectral variance analysis (SVA) to identify periodic fluctuations, high-pass filtering (HPF) and filtering with Fourier transform smoothing (FTS). HPF is used to remove long periodic fluctuations (low frequencies) up to a distinct cutoff period that is chosen according to the SVA (see section 3.3), the procedure results in a loss of data at the beginning and the end of the data series (as shorter the cutoff period, as greater is the loss of data!). This disadvantage is overcome by filtering with FTS. But because the cutoff frequency in the FTS procedure (see section 3.4) can only be a frequency of one of the harmonics of the data series (Jenkins and Watts, 1968), the actual cutoff frequency and the corresponding frequency of maximal variance power (= mean period length of the fluctuation) in the SVA may be differing; this can lead to an incorrect transformation, especially in the case of long periodic fluctuations (low frequencies). Therefore, FTS is mostly used to remove fluctuations of higher frequencies. The stepwise elimination of deterministic variability components (Figure 1, approach (1)) results in a stepwise variability reduction of the remaining data series. At the final stage, the residuals contain the stochastic variability component according to the constraints given above.

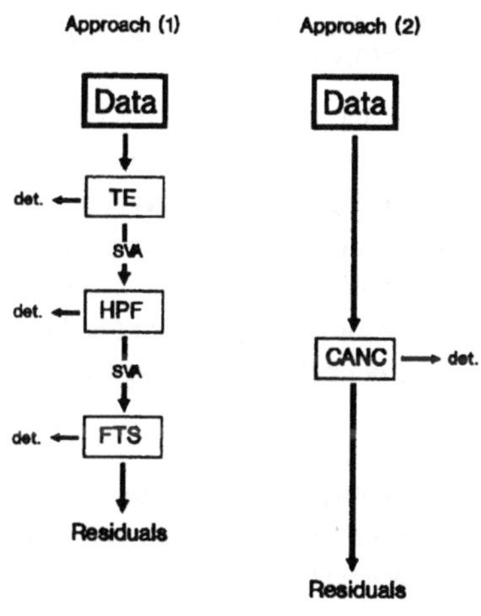

Figure 1: Scheme of separation of deterministic and stochastic variability components.

In the *second approach* (section 3.5), a filter technique called "Correlated Data Adaptive Noise Cancelling" (CANC) (Hattingh, 1988) is used to separate deterministic and stochastic variability components of the space series of groundwater solute concentrations. The CANC method provides a least-square estimate of the *spatially predictable* variability component, i.e. of the deterministic variability according to the definition given in section 1. After elimination of the de-

terministic variability component with the CANC method, the residuals contain only stochastic variability.
Because deterministic variability is defined more stringent in approach (1) (spatially predictable *and* related causally to a known factor) than in approach (2) (spatially predictable only), this portion calculated by the different mathematical methods of approach (1) or (2), respectively, may be differing.
Notice that the variability of the residuals calculated by approach (1) or (2) (variance or standard deviation, respectively) can only be a biased estimate of the true variability, because a calculated trend function (e.g. polynomial) or any filter function f*(x) is only an estimate of the true trend or filter function f(x). Although the expectation of [f(x) - f*(x)] equals zero (because f*(x) is an unbiased estimate of f(x)), the variable [f(x) - f*(x)] has a finite, but unknown variance.
The mathematical and statistical methods used are briefly outlined below.

3.1 Trend verification
To verify the presence of a trend in the data series we use the method of Davidoff et al. (1986). The drift function is given by

$$D(h,C) = \frac{1}{N(h,C)} \sum_{i=1}^{N(h,C)} \left(Xc_i - Xc_{i+h}\right) \quad (1)$$

where $(Xc_i - Xc_{i+h})$ is the ith pair of data separated by the lag h in the complete set C, and N(h,C) is the total number of pairs in C. A complete set C is constructed by a random selection of nonadjacent pairs of observations of lag h "...such that if an observation is adjacent to an observation in a previously selected pair, then the common observation from the new pair will be excluded from the remaining observations" (Davidoff et al., 1986). A significance level for the drift function is calculated using a t statistic. If a significant trend is present in a data series, it is eliminated by fitting a first or second order polynomial using least squares.

3.2 Spectral variance analysis
Spectral variance analysis is used to identify periodic fluctuations in the data series. The power spectrum, which is a Fourier transform of the autocorrelation function, is calculated for a given frequency f according to Jenkins and Watts (1968) by

$$R(f) = 2\left\{1 + \left[2 \sum_{h=1}^{L-1} r(h)\, w(h) \cos(2\,\pi\, f\, h)\right]\right\} \quad (2)$$

where r(h) is the autocorrelation at lag h (calculated according to Schönwiese, 1985), w(h) is a weighting function, called lag window (we use a Tukey lag window, Jenkins and Watts, 1968), and L is the maximum lag h (truncation point). Significant peaks in the power spectrum indicate frequencies or periods ($p=f^{-1}$) of pronounced periodicities in the data series. The significance of peaks is tested against the hypothesis "the power spectrum represents the spectrum of a purely random process (white noise)" by calculating the theoretical spectrum of a white noise process and its confidence limits according to Schönwiese (1985). Peaks exceeding the 90% or 95% confidence limit, respectively, are considered to be significant.

3.3 High-pass filtering

To remove long periodic fluctuations from space series of groundwater solute concentration data, which can be assigned to extrinsic factors, we use a high-pass filter (Schönwiese, 1985). The vector of high-pass filtered data [â (**H**)] is calculated by

$$[\hat{a}(\mathbf{H})] = [\mathbf{X}] - [\hat{a}(\mathbf{L})] \tag{3}$$

where [**X**] is the vector of input data to be high-pass filtered and [**â**(**L**)] is the vector of low-pass filtered data. The low-pass filtering operation is done by multiplying subsets of the input data with a set of filter weights

$$\hat{a}(L)_{(j)} = [\mathbf{X}]_{(j)}\,[\mathbf{W}']^T \tag{4}$$

where j = 1, ..., N-2m.

In equation (4), $\hat{a}(L)_{(j)}$ is the jth low-pass filtered value, N is the length of the input data series,

$$[\mathbf{X}]_{(j)} = \left[x_{i-m}, \ldots, x_{i-0}, \ldots, x_{i+m}\right] \tag{5}$$

is the jth subset of data to be filtered and [**W'**] is the vector of filter weights. The length of the filter equals 2m+1, therefore the filter is symmetric, and so, no phase shift can occur in the filtered data. The rough filter weights are calculated by

$$w_k = \frac{1}{\sqrt{2\pi}} \exp\left[-0.5\,(k\,6\,p_0^{-1})^2\right] \tag{6}$$

where p_0 is the cutoff period and k=0,1,2,...,m (Schönwiese, 1985). Theoretically, an infinite number of filter weights is calculated by eq. (6). But for practical purpose, the number of filter weights is restricted to the condition $w_m \leq w_0/10$, and because the filter is symmetric, $w_{-k} = w_k$. Finally, the rough filter weights are normalized, so that

$$\sum_{k=-m}^{m} w'_k = 1 \tag{7}$$

where w'_k are the normalized filter weights. Because eq. (6) is related to the Gaussian normal distribution, the filter technique is called "Gaussian low-pass filtering" (Schönwiese, 1985).

3.4 Fourier transform smoothing

Fluctuations of higher frequencies of the groundwater solute concentrations that can be directly compared to the spatial patterns of well known extrinsic influences (e.g. pattern of overlapping fertilizer strips on arable land, pattern of canopy throughfall in coniferous forest), are eliminated by Fourier transform smoothing. This method was suggested by Kimball (1974) to smooth agronomic data. The Fourier coefficients of the data series are calculated for distinct frequencies $f = n/(N\,\delta)$ according to Jenkins and Watts (1968) by

$$A(f) = \delta \sum_{k=0}^{N-1} x'(k) \cos\left[(2 \pi k n)/(N \delta)\right] \tag{8}$$

$$B(f) = \delta \sum_{k=0}^{N-1} x'(k) \sin\left[(2 \pi k n)/(N \delta)\right] \tag{9}$$

where n = -N/2,..., 0, ..., (N/2)-1.

In equations (8) and (9), δ represents the sampling interval, x'(k) is the kth value of a converted data series, which is calculated by conversion of the original data using the translation-rotation algorithm of Hayes et al. (1973), and N is the number of data. From the Fourier coefficients a variance line spectrum is calculated (Kimball, 1974). Then the data are smoothed by an inverse Fourier transform with a restricted number of Fourier coefficients. For this purpose, all Fourier coefficients of frequencies exceeding a selected cutoff frequency f_0 are set equal to zero. The selection of the cutoff frequency from the line spectrum must be in correspondence with the results of the power spectrum calculation (Böttcher and Strebel, 1988a). Finally, the smoothed data are restored by an inverse translation-rotation.

3.5 CANC method

The CANC algorithm was recently published by Hattingh (1988) as a method to remove noise (stochastic variability) from geophysical data series. To apply the CANC method, a primary signal X (a data series of N equally spaced values x_i) and a reference signal Rs are needed as filter input. The reference signal can be generated from the primary signal by delaying it in time/space and adding some random noise (Hattingh, 1988). We use the following equation:

$$rs_i = \left(x_{i+1}\ 0.5\right) + \left\{x_{i+1}\ (RND - 0.5) / 10\right\} \tag{10}$$

rs_i is the ith reference signal (i = 1, ... , N) and RND is an equally distributed random number (0 to 0.999...). To generate rs_N, x_1 is used. Own test calculations (manuscript in preparation) showed, that the above equation is well suited to optimize the results of the CANC method.

The filtering operation is done by

$$y_i = [\mathbf{Rs}]_{(i)} [\mathbf{F}]^T_{(i)} \tag{11}$$

where $L \leq i \leq N$, and L is the filter length (according to Hattingh, 1988, the filter length is set to L = 8).

In eq. (11), y_i is the ith filter output, $[\mathbf{F}]_{(i)}$ is the ith vector of filter coefficients (L elements; the initial filter coefficients are set to zero), and the vector

$$[\mathbf{Rs}]_{(i)} = \left[rs_i, rs_{i-1}, \ldots, rs_{i-L+1}\right] \tag{12}$$

is the ith subset of a delayed version of the reference signal (L elements, delay must equal L). The filter output y is then substracted from the primary signal, which must be delayed by one-half of the filter length, to give the error signal e

$$e_i = x_{i-L/2} - y_i \tag{13}$$

This error signal is used in the Widrow-Hoff algorithm (Hattingh, 1988) to adjust the filter coefficients

$$[\mathbf{F}]_{(i+1)} = [\mathbf{F}]_{(i)} + \left(2\,\mu\,e_i\right)[\mathbf{Rs}]_{(i)} \tag{14}$$

where μ is a parameter determining the rate of convergence and stability of the filter. The filtering and adjusting operation is repeated iteratively for each data point of the data series until the last point has been reached and then started again. This process is repeated iteratively until it converges. The filter output then is a best least-square estimate of the deterministic component of the primary signal. For details of the theoretical background and the practical application of the CANC method, the reader is referred to Hattingh (1988).

4. Results and discussion

The presentation of results is confined to two examples of particularly important groundwater solute concentrations. The concentration distributions of nitrate measured along the transect beneath arable land, and of sulfate measured along the transect beneath coniferous forest are shown in Figure 2. The data are evaluated according to approach (1) and (2) outlined in section 3.

The *nitrate concentrations* under arable land (Figure 2a), determined perpendicular to the direction of cultivation, show regular periodic fluctuation. This is a typical example of a concentration pattern caused by overlapping of the strips of fertilizer application.

Approach (1): The calculation of the drift function (eq. 1) and the t statistic (not depicted) revealed no significant trend in the nitrate concentration data. The periodic fluctuation is confirmed by a significant peak at $f = 0.080\ m^{-1}$ in the corresponding power spectrum (Figure 3a), representing a period length of $p \approx 12$ m. The period length of this regular fluctuation of nitrate concentrations at the groundwater table under arable land results from the width of the strip covered by one pass of the fertilizer spreader, which is about 12 m (Böttcher and Strebel, 1988b). This deterministic variability component is eliminated by Fourier transform smoothing, using a cutoff frequency f_0 of $0.066\ m^{-1}$ (not depicted). The spectral variance analysis of the residuals after FTS shows now a significant peak at a frequency of $f = 0.173\ m^{-1}$ (Figure 3b). This periodic fluctuation of $p \approx 6$ m forms only an insignificant peak in the power spectrum of the initial data (Figure 3a), because the low frequency fluctuation of $f = 0.080\ m^{-1}$ is the dominant contribution to the total variance. Since the periodicity of 6 m period length can also be ascribed to the fertilization practice (overlapping of fertilizer strips of former application, 180° out-of-phase), it has to be considered as a deterministic variability component, and is therefore eliminated by FTS using a cutoff frequency $f_0 = 0.197\ m^{-1}$. After this second FTS step the residuals of the nitrate concentration data still contain significant fluctuations of higher frequencies (power spectrum not depicted). These fluctuations cannot be assigned to any known extrinsic or intrinsic factor and are thus not removed. The stepwise variability reduction (in terms of % of total standard deviation) is depicted in Figure 5a. Because the data reveal no trend and no long periodic fluctuations, the only decrease of standard deviation σ is due to Fourier transform smoothing (decrease of σ from 100% to 32% in two steps).

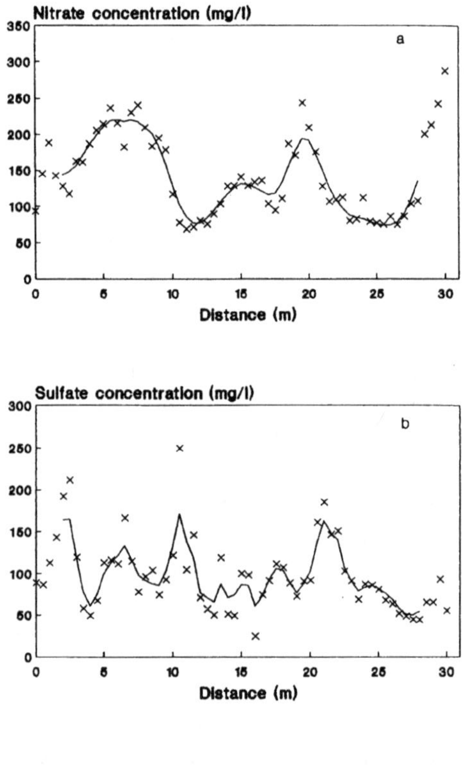

Figure 2: Groundwater concentrations (crosses) of nitrate under arable land (a), and of sulfate under coniferous forest (b), respectively, at the water table along a 30 m transect. The solid line indicates the deterministic variability component separated by the CANC method.

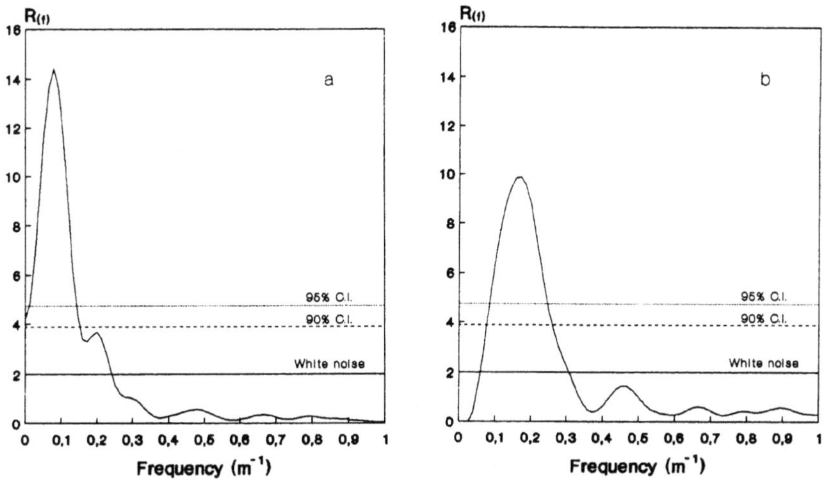

Figure 3: Power spectra for the nitrate concentrations plotted in Figure 2a: a) for the initial data; b) for the residuals after FTS with cutoff frequency $f_0 = 0.066\ m^{-1}$

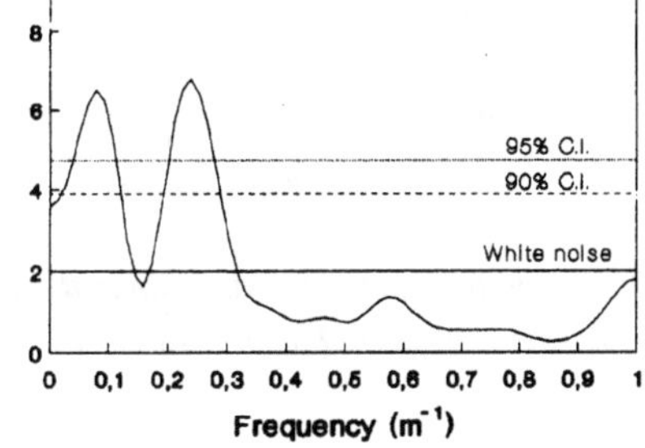

Figure 4: Power spectrum for the sulfate concentrations plotted in Figure 2b.

Approach (2): Filtering of the nitrate concentration data by the CANC method results in the curve given in Figure 2a as solid line. The standard deviation σ is decreased from 100% to 29% (Figure 5a). The results of approach (1) and (2) are more or less the same. So, the variability components that are spatially predictable by the CANC method, correspond to those that can be assigned to extrinsic factors by the methods used in approach (1). As the residuals contain only the stochastic variability component, the contribution of deterministic variability components to the total variability of groundwater nitrate concentrations under arable land is in our example in the order of 70% of the total standard deviation.
The distribution of *sulfate concentrations* at the groundwater table along the transect beneath coniferous forest is shown in Figure 2b. The variability pattern of the sulfate concentrations is more or less erratic.

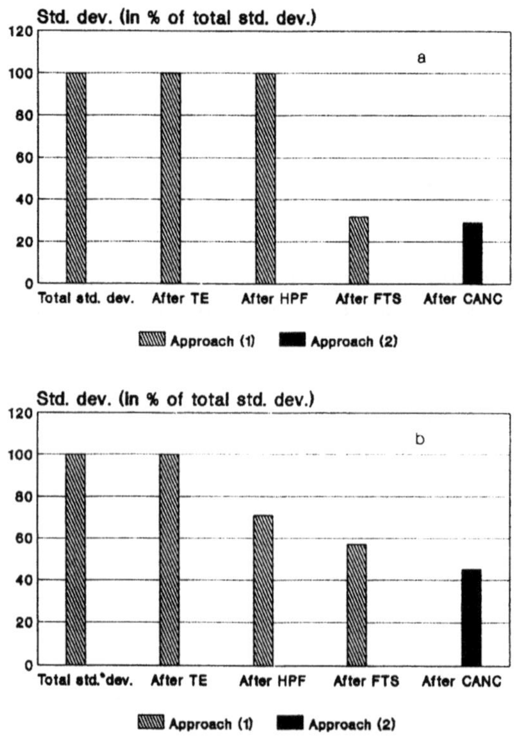

Figure 5: Decrease of variability due to stepwise elimination of deterministic variability components: a) nitrate concentrations; b) sulfate concentrations.

Approach (1): Trend verification revealed no significant trend in the data. Figure 4b represents the power spectrum, calculated using equation (2). It shows two significant peaks, indicating that a long periodic fluctuation ($p \approx 15$ m) is superposed by a short periodic fluctuation of period length $p \approx 4$ m. The long periodic fluctuation is very probably caused by different canopy densities of the pine stand (these differences may be randomly distributed within the pine stand, but they determine a regular fluctuation along the sampled transect). This deterministic variability component is eliminated using the high-pass filtering technique with a cutoff period p_0 of 6.25 m, and calculated normalized filter weights $w'_0 = 0.193$, $w'_1 = 0.172$, $w'_2 = 0.122$, $w'_3 = 0.068$, $w'_4 = 0.031$ and $w'_5 = 0.011$ (notice that $w'_{-k} = w'_k$). The short periodic fluctuation in the sulfate concentration data ($p \approx 4$ m) is caused by the interception deposition of atmospheric constituents at the canopy (especially SO_2) and throughfall, which preferentially drips at the edges of the tree canopies onto the ground and infiltrates into the soil (Böttcher and Strebel, 1988c). Using Fourier transform smoothing with a cutoff frequency f_0 of 0.275 m^{-1}, this deterministic variability component was removed from the data series. Now the residuals contain no more significant fluctuations (power spectrum not depicted). The variability decrease by applying approach (1) is shown in Figure 5b. The standard deviation (in terms of % of the total standard deviation) is decreased in two steps from 100% to 57%.

Approach (2): The solid line in Figure 2b indicates the variability component that is spatially predictable by the CANC method. The elimination of deterministic variability components using the CANC method decreases σ from 100% to 45%. The difference between the results of approach (1) and (2) is probably not significant, because as yet unpublished test calculations (manuscript in preparation) indicate, that the CANC method tends to overestimate the portion of deterministic variability, if the ratio of σ-deterministic: σ-stochastic is < 2. The contribution of deterministic variability to the total σ of the sulfate concentrations under coniferous forest is in our example in the range of 40% to 50% and thus less pronounced then in the case of the nitrate concentrations under arable land.

5. Conclusions

Deterministic and stochastic variability components of equally spaced data series can be separated and quantified using two different approaches (time/space series analysis, CANC filter method). The results presented indicate, that the "local scale" spatial variability of transect data of solute concentrations at the groundwater table beneath arable land and coniferous forest on sandy soils can be strongly influenced by deterministic, spatially predictable variability components. The causes for these deterministic variability patterns are mostly extrinsic factors, i.e., patterns of solute input into the soil (overlapping of strips of fertilizer application, canopy interception deposition and throughfall) can be preserved during solute transport through the unsaturated soil and thus reflected at the groundwater table. The stochastic variability components may be caused by sampling errors, analytical errors, randomly distributed, inhomogeneous solute input into the soil, and randomly distributed, heterogeneous water flow and solute transport in the soil.

The values of deterministic and stochastic variability given in this paper are only estimates of the true values, because the calculated variance of the residuals is not unbiased (see section 3). Further work is needed to evaluate this unavoidable estimation error (a manuscript on test calculations of the estimation error of the CANC method is in preparation).

Acknowledgements

We thank the Deutsche Forschungsgemeinschaft and the Umweltbundesamt for financial support of this study.

References

Böttcher, J. and O. Strebel, 1988a: *Spatial variability of groundwater solute concentrations at the water table under arable land and coniferous forest. Part 1: Methods for quantifying spatial variability (geostatistics, time series analysis, Fourier transform smoothing).* Z. Pflanzenernähr. Bodenk,., **151**, 185-190.

Böttcher, J. and O. Strebel, 1988b: *Spatial variability of groundwater solute concentrations at the water table under arable land and coniferous forest. Part 2: Field data for arable land and statistical analysis.* Z. Pflanzenernähr. Bodenk., **151**, 191-195.

Böttcher, J. and O. Strebel, 1988c: *Spatial variability of groundwater solute concentrations at the water table under arable land and coniferous forest. Part 3: Field data for a coniferous forest and statistical analysis.* Z. Pflanzenernähr. Bodenk., **151**, 197-203.

Davidoff, B., J.W. Lewis, and H.M. Selim, 1986: *A method to verify the presence of a trend in studying spatial variability of soil temperature.* Soil Sci. Soc. Am. J., **50**, 1122-1127.

Hattingh, M., 1988: *A new data adaptive filtering program to remove noise from geophysical time- or space-series data.* Computers and Geosciences, **14**, 467-480.

Hayes, J.W., D.E. Glover, and D.E. Smith, 1973: *Some observations on digital smoothing of electroanalytical data based on the Fourier Transformation.* Anal. Chem., **45**, 277-284.

Jenkins, G.M., and D.G. Watts, 1968: *Spectral analysis and its applications.* Holden Day, San Francisco, 525 pp.

Kimball, B.A., 1974: *Smoothing data with Fourier transformation.* Agron. J., **66**, 259-262.

Russo, D., and W.A. Jury, 1987: *A theoretical study of the estimation of the correlation scale in spatially variable fields. 2. Nonstationary fields.* Water Resour. Res., **23**, 1269-1279.

Schönwiese, Chr.-D., 1985: *Praktische Statistik für Meteorologen und Geowissenschaftler.* Gebrüder Borntraeger, Berlin, Stuttgart, 231 pp.

Strebel, O. and J. Böttcher, 1989: *Solute input into groundwater from sandy soils under arable land and coniferous forest: Determination of area-representative mean values of concentration.* Agric. Water Manage., **15**, 265-278.

J. Böttcher and O. Strebel, Bundesanstalt für Geowissenschaften und Rohstoffe, Stilleweg 2, D-3000 Hannover 51, Fed. Republ. of Germany.

Field-Scale Water and
Solute Flux in Soils
Monte Verità

USE OF SCALING TECHNIQUES TO QUANTIFY VARIABILITY IN HYDRAULIC FUNCTIONS OF SOILS IN THE NETHERLANDS

J.H.M. Wösten

Quantification of variability in hydraulic functions of different soils allowed calculation of the variability in output of models which use these functions as input. Variability of an output variable defines the uncertainty of a particular calculation and is therefore indispensable for applications which need a measure of accuracy. Meaningful description of variability requires a minimum number of measured hydraulic functions. Defined accuracies of calculated functional criteria which are practical aspects of soil behavior, were used as references to classify soils, for which insufficient hydraulic functions are known in larger soil groups. For the Netherlands this classification resulted in three large soil groups; coarse-textured, medium-textured and fine-textured, each comprising at least 30 measured functions. Scaling was successfully used to reduce the variation in measured hydraulic functions into a narrow band around the scaled mean hydraulic function for each soil group. Distribution functions of scale factors were used to transform the scaled mean hydraulic conductivity and moisture retention functions into 100 new hydraulic functions for each soil group. Variability in measured and in newly generated functions was compared on the basis of calculated functional criteria. Mean values of functional criteria are close for the measured and newly generated sets of functions. The percentage of variation in functional criteria explained by the newly generated hydraulic functions varied from 15% to 96%. Although scaling is an attractive technique to simplify description of the complex hydraulic heterogeneity of soils in the Netherlands, it underestimates the variability.

1. INTRODUCTION

Soil water retention and hydraulic conductivity functions are crucial ingredients for the estimation of rates of soil water flow. During the past decade hydraulic functions have been measured for a wide variety of soils in The Netherlands. The 197 measured functions were classified into 20 different soil groups according to soil texture (as used by the former Netherlands Soil Survey Institute), and type of horizon, being either topsoil (A horizon) or subsoil (B and C horizons). Tabulated forms of the geometrically averaged functions for the 20 soil groups were presented by Wösten et al. (1987). These average functions facilitate soil physical interpretations of soil maps which are relevant for land evaluation and environmental and hydrological studies that require soil physical information for the unsaturated zone of areas of land (e.g. Wösten et al., 1985; Breeuwsma et al., 1986).

Individual functions within each of the 20 soil groups show considerable variability due to differences in soil texture, organic matter content and bulk density. Similar variability in hydraulic functions measured in the field is reported by other investigators (e.g. Nielsen et al., 1973; Hopmans, 1987). Until now, existing variability in hydraulic functions of soil groups distinguished in the Netherlands has not been quantified. Consequently model output, so far, does not reflect the effects of this variability. Presentation of model output in the form of a mean and variance becomes increasingly attractive because it reflects and quantifies the uncertainty of a particular calculation. In order to provide this measure of accuracy of model output, quantification of variability in hydraulic functions is a necessity (e.g. Bresler and Dagan, 1979; Van der Zee and Van Riemsdijk, 1987).

Quantification of variability in hydraulic functions of a soil group is only meaningful if a minimum number of functions is measured for that particular soil group. Warrick et al. (1977a) showed that the accuracy of estimated mean soil water fluxes strongly increased with increasing number of samples. If in turn a certain accuracy in calculated functional criteria is taken as a reference, the number of samples can be determined. Soil groups containing insufficient samples are combined into larger, new groups.

Once sufficiently large soil groups are established, variability in hydraulic functions is quantified with the scaling theory of similar media. The theory of scaling (Miller and Miller, 1955, 1956) assumes that similar media differ only in the scale of their internal microscopic geometries, and therefore have equal porosities. Scaling relates the hydraulic functions $h(\theta)$ and $K(\theta)$ for soils at different locations to a representative mean through a single stochastic variable, the scale factor. The complexity of variability in hydraulic functions is thus simplified to the description of distribution functions of scale factors. Methods to determine scale factors are described by Warrick et al. (1977b), Simmons et al. (1979) and Russo and Bresler (1980). Sharma and Luxmoore (1979) and Clapp et al. (1983) used scaling techniques to investigate the practical consequences of soil heterogeneity on water budget components in a watershed. Similar to studies by Ahuja et al. (1984), and Hopmans and Stricker (1989), distribution functions of scale factors are used, in this study, to generate a new set of hydraulic functions for each soil group. Variability in measured and newly generated hydraulic functions is compared by comparing calculated functional criteria which relate to practical applications.

The specific objectives of this study were hence (1) to form soil groups that contained sufficient individual hydraulic functions to allow quantification of variability in these functions, (2) to apply scaling as a technique for quantification of variability within each soil group, and (3) to examine the effectiveness of scaling in describing variability in measured hydraulic data.

2. MATERIALS AND METHODS

A total of 197 water retention and hydraulic conductivity functions were measured for different soils in the Netherlands. The measured functions for the soils, which were classified in 20 different soil groups, form a unique data base covering a broad spectrum of soils.
Hydraulic conductivities were measured using a combination of the following five methods :

(1) The column method (e.g., Bouma, 1982) for the vertical saturated hydraulic conductivity, K_S.
(2) The crust-test (Bouma et al., 1983) for unsaturated conductivities when the pressure head, h, is between 0 and -50 cm.
(3) The sorptivity-method (Dirksen, 1979) for conductivities of coarse-textured soils when h < -50 cm.
(4) The hot-air method (Arya et al., 1975) for conductivities of medium- and fine-textured soils when h < -50 cm.
(5) The evaporation method (Boels et al., 1978) for hydraulic conductivities when h is between 0 and -800 cm.

Soil water retention functions were obtained by slow evaporation of wet, undisturbed samples in the laboratory as reported by Boels et al. (1978) and Bouma et al. (1983). In this method, pressure heads were periodically measured with transducer-tensiometers while at the same time subsamples were taken to determine water contents, thus yielding points relating to h and θ. Water contents corresponding with pressure heads lower than -800 cm of water were obtained by conventional methods using air pressure (Richards, 1965). For relatively fine-textured soils, a staining technique was applied to record the effects of horizontal cracks on the upward flux of water from the water table to the rootzone (Bouma, 1984).
The measured water retention and hydraulic conductivity functions were fitted in a previous study (Wösten and Van Genuchten, 1988) with the following closed-form analytical expressions of Van Genuchten (1980):

$$S = \frac{\theta - \theta_r}{\theta_s - \theta_r} = \left[1 + |\alpha h|^n\right]^{-m} \tag{1}$$

and

$$K(S) = K_s S^{\iota}\left[1 - \left(1 - S^{1/m}\right)^m\right]^2 \tag{2}$$

The parameter S is the effective saturation. The subscripts r and s refer to residual and saturated values of the volumetric water content θ. K_S is the saturated hydraulic conductivity and α, n, m and ι are parameters which determine the shape of the functions. A modified form of a nonlinear least squares optimization program developed by van Genuchten (1978) was used to estimate the unknown parameters (θ_r, θ_S, α, n, ι and K_S) simultaneously from measured water retention and hydraulic conductivity data. Following Wösten and Van Genuchten (1988), the parameter θ_r was fixed at zero and the parameters θ_s and K_s were fixed at their independently measured values. Consequently, only three unknown parameters (α, n and ι) were needed to fully describe the individual functions. In this study the fitted hydraulic functions were used.

3. DATA ANALYSIS

The minimum number of functions required for a meaningful description of variability in each soil group was estimated using the data of the largest of the presently distinguished 20 soil groups. The selected soil group comprised 49 measured functions and had the following ranges in soil properties: silt 1-9% ; organic matter 0.1-2% ; bulk density 1.4-1.8 g cm^{-3}. The 49 functions were used to calculate three functional criteria which are practical aspects of soil behavior that have immediate relevance to management-type applications (Wösten et al., 1986):

(1) Travel time (T) of water from the soil surface down to a water table assumed to be at a depth (D) of 1 m. The travel time is approximated by $T = \theta D/q_d$, where q_d is the average daily vertical downward flux for Dutch winter conditions (0.14 cm d^{-1}), and θ is the profile-averaged water content corresponding to a flux q_d. θ is derived from the K-θ function.

(2) Depth of water table (L) which can sustain a given upward flux of water to the soil surface or to the bottom of the root zone. As shown by Gardner (1958), this depth may be obtained by integrating Darcy's law to give:

$$L = \int_0^h \frac{1}{1 + q_u/K(h)} dh \qquad (3)$$

where h is the imposed soil water pressure head (arbitrarily taken as -500 cm) at the bottom of the root zone, and q_u is the steady-state upward water flux density, assumed to be 0.2 cm d^{-1}.

(3) Downward flux of water (F) corresponding to the minimum soil air content, θ_a (taken to be 5%) which is required to maintain adequate aeration in the root zone for maximum root activity and crop growth (FAO, 1985). Downward flux of water follows immediately from the retention and conductivity function at a water content which is 0.05 cm^3cm^{-3} less than the saturated value.

In this study emphasis was placed on the accuracy of predicting new values of the functional criteria rather than on the accuracy of predicting mean values. Accuracy of predicting new values, as expressed by the Standard Error of Prediction (SEP), depends on the standard deviation (sd) of the functional criterion and the sample size (n), according to:

$$SEP = sd \sqrt{1 + 1/n}$$

SEP (Montgomery and Peck, 1982) values are used to establish the relation between the width of the 90% prediction interval and sample size. This width indicates that new values of the functional criteria are with 90% certainty within the limits indicated by the width. In turn, the relation between width and sample size can be used to determine the minimum number of samples required to obtain a defined accuracy.

Combination of soil groups containing insufficient hydraulic functions is based on an evaluation of the discriminative power of the two soil properties; "soil texture and type of horizon" presently used to classify the 20 soil groups. Only the most discriminative soil property (e.g. soil texture or type of horizon) was used to re-classify the insufficiently large soil groups into larger, new groups.

Variability in hydraulic functions of the sufficiently large soil groups was quantified with scaling. According to scaling theory the dimensionless scale factor, a_r, is defined as the ratio of the microscopic characteristic length l_r of a soil at location r and the characteristic length l_m of a reference soil:

$$a_r = l_r / l_m \tag{5}$$

If a reference soil water pressure head h_m and conductivity K_m together with the scale factors are known, the water pressure head (h_r) and hydraulic conductivity (K_r) at given water contents at any location r is calculated according to:

$$h_r = h_m / a_r \tag{6}$$

and

$$K_r = K_m a_r^2 \tag{7}$$

Because soils do not have identical porosities, h and K are written as functions of the effective saturation S (Eq. 1) rather than functions of the volumetric water content θ. The saturated hydraulic conductivity K_s is not included in the scaling procedure because it is often controlled by flow in macropores, whereas the unsaturated hydraulic conductivity is controlled by the entire continuum of pore sizes (Jury et al., 1987a). Specific scaling methods differ according to how the reference $h_m(S)$ and $K_m(S)$ functions are determined. Hopmans (1987) concluded from a comparison of five different scaling methods that the method described by Warrick et al. (1977b) was the most promising. Therefore, only the latter scaling method was used in this study.

After establishing the distribution function of the calculated scale factors for each soil group, this distribution function was used, in turn, to generate randomly a set of 100 new scale factors. Next, the new scale factors were used to transform the or reference soil water retention and hydraulic conductivity functions into 100 new hydraulic functions according to Eqns. (6) and (7). Variability in these newly generated hydraulic functions was analysed on the basis of the variability of the calculated functional criteria described earlier. Comparison of the means and variances of functional criteria calculated from the newly generated and the measured hydraulic functions allowed an evaluation of the effectiveness of scaling in describing variability in soil hydraulic data.

When functional criteria and scale factors were normally distributed, the mean (m_1) and variance (sd_1^2) of these criteria and factors were estimated using untransformed data according to:

$$m_1 = \frac{1}{n}\sum_{i=1}^{n} x_i \tag{8}$$

and

$$sd_1^2 = \frac{1}{n-1}\sum_{i=1}^{n} (x_i - m_1)^2 \tag{9}$$

In these equations x_i is the ith observation and n is the sample size. However, if these same criteria were lognormally distributed, then their mean (m_2) and variance ($sd_2{}^2$) was estimated using transformed data according to:

$$m_2 = \exp(\mu + \sigma^2/2) \tag{10}$$

and

$$sd_2^2 = \exp(2\mu + \sigma^2)[\exp(\sigma^2) - 1] \tag{11}$$

where

$$\mu = \frac{1}{n}\sum_{i=1}^{n} \ln(x_i) \tag{12}$$

and

$$\sigma^2 = \frac{1}{n-1}\sum_{i=1}^{n} (\ln(x_i) - \mu)^2 \tag{13}$$

The coëfficient of variation (CV) was calculated by dividing the standard deviation (sd) by the mean (m).

4. RESULTS AND DISCUSSION

Relations between half widths of 90% prediction intervals for three functional criteria and sample sizes are shown in Figure 1 for the soil group comprising 49 measured functions. The figure indicates that the accuracy of predicting the functional criteria, as expressed by the half width of the 90% prediction interval, increases with increasing sample size. In this study a difference of 5% between the half width of the 90% prediction interval and the true half width (where SEP $\rightarrow$ sd when n $\rightarrow \infty$) is used as a reference. Figure 1 indicates that, in order to meet this requirement, soil groups should comprise about 30 functions.

Linear regression of calculated functional criteria versus soil texture and type of horizon, for all 197 individual functions, showed that the discriminative power of the soil property; "soil texture" is large compared to the discriminative power of the soil property; "type of horizon". This result of the regression analysis combined with a required sample size of at least 30 functions led

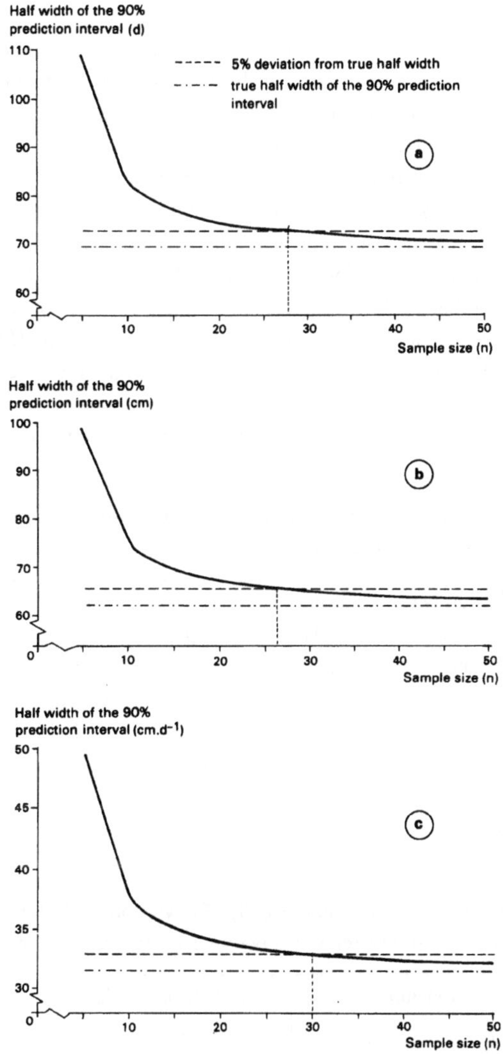

Figure 1. Half width of the 90% prediction interval as a function of the sample size for the functional criteria: a) travel time, b) depth of water table and c) downward flux of water. For the calculation of these data the used soil group comprised 49 measured functions which had the following ranges in soil properties: silt 1-9%; organic matter 0.1-2%; bulk density 1.4-1.8 g cm^{-3}

to consolidation of the original, distinguished 20 soil groups into 3 new, sufficiently large groups; coarse-textured, medium-textured and fine-textured soils. The number of functions and ranges in soil texture, organic matter content, bulk density and median of the sand fraction (M50) for each of the three new soil groups are presented in Table 1.

Figures 2, 3 and 4 show the unscaled and scaled water retention data and hydraulic conductivity data for the coarse-textured, medium-textured and fine-textured soils, respectively. Scaled data resemble the characteristic shapes of hydraulic functions normally measured for sandy, loamy and clay type soils. Percentage reductions in the Sum of Squares (SS) of deviations between scaled mean hydraulic function and individual hydraulic data for each soil group, before and after scaling, are presented in Table 2. A high reduction in SS indicates that scaling was successful. Table 2 and Figures 2, 3 and 4 illustrate that scaling was indeed effective in reducing the scatter of the original data points into a narrow band around the mean scaled functions. Effectiveness was greater for water retention as compared to conductivity data. Because K_s was not included in the scaling procedure, its variation did not decrease after scaling.

Table 1. Number of individual functions, ranges in soil texture, organic matter content, bulk density and M50 for three, newly generated soil groups.

soil group	number of functions	silt, 2-50 μm (%)	clay, >2μm (%)	organic matter (%)	bulk density (g/cm^3)	M50[1] (μm)
coarse-textured	105	4-49		0.1-13	1.1.-1.8	130-180
medium-textured	43		9-22	0.2-6	1.2-1.7	
fine-textured	49		26-77	0.1-15	0.9-1.7	

[1] median sand fraction

Table 2. Percent reduction in sum of squares (SS) of deviations between scaled mean hydraulic functions and individual hydraulic data points before and after scaling, for three soil groups.

soil group	reduction in SS (%) h-values	
coarse-textured	80	53
medium-textured	93	35
fine-texutred	90	27

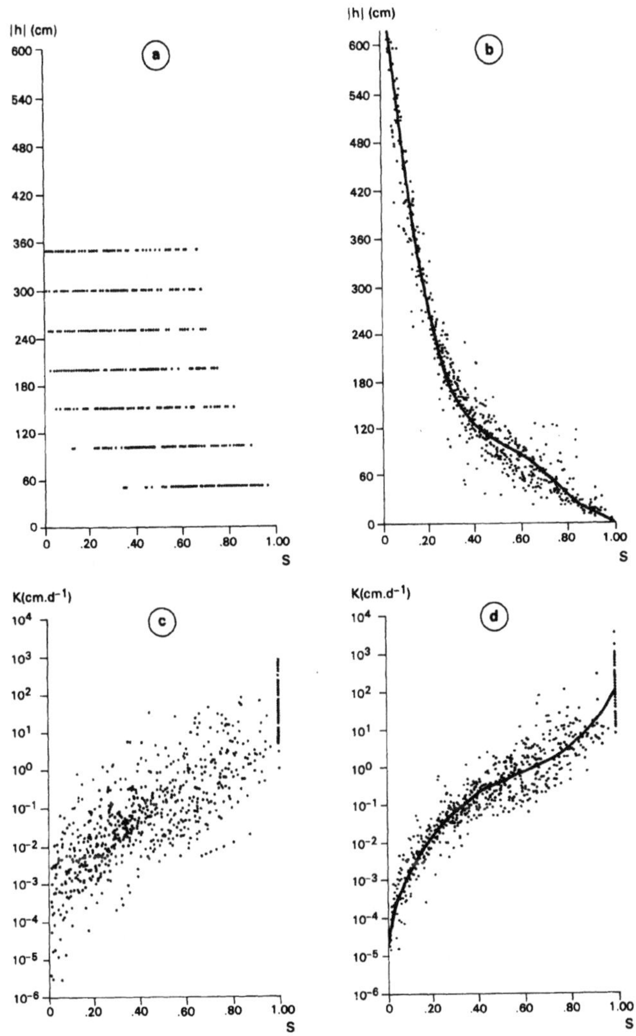

Figure 2. Unscaled (a) and scaled (b) water retention data and unscaled (c) and scaled (d) hydraulic conductivity data for the coarse-textured soil group.

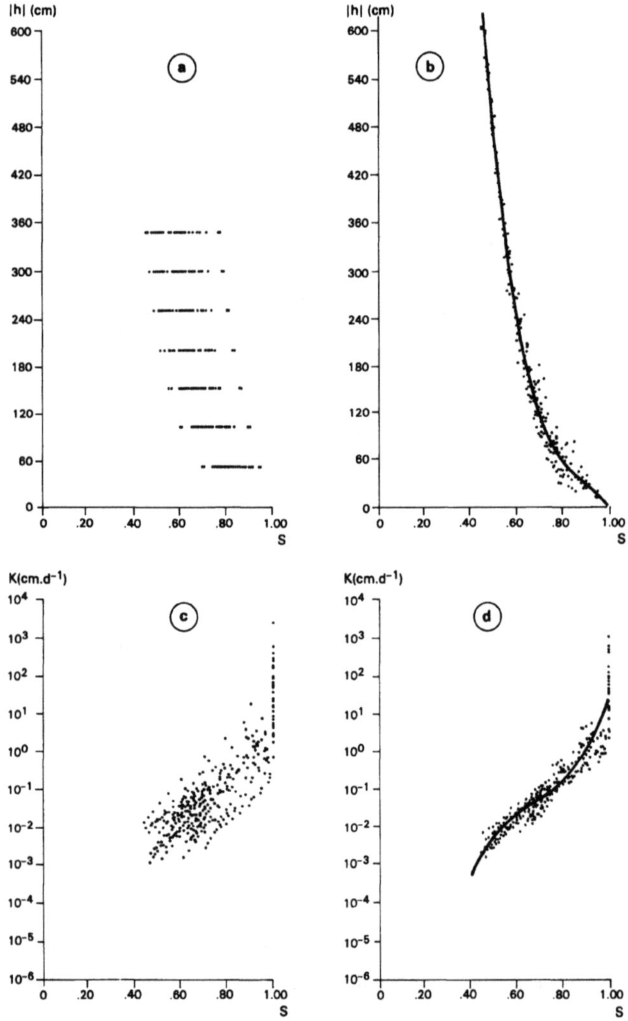

Figure 3. Unscaled (a) and scaled (b) water retention data and unscaled (c) and scaled (d) hydraulic conductivity data for the medium-textured soil group.

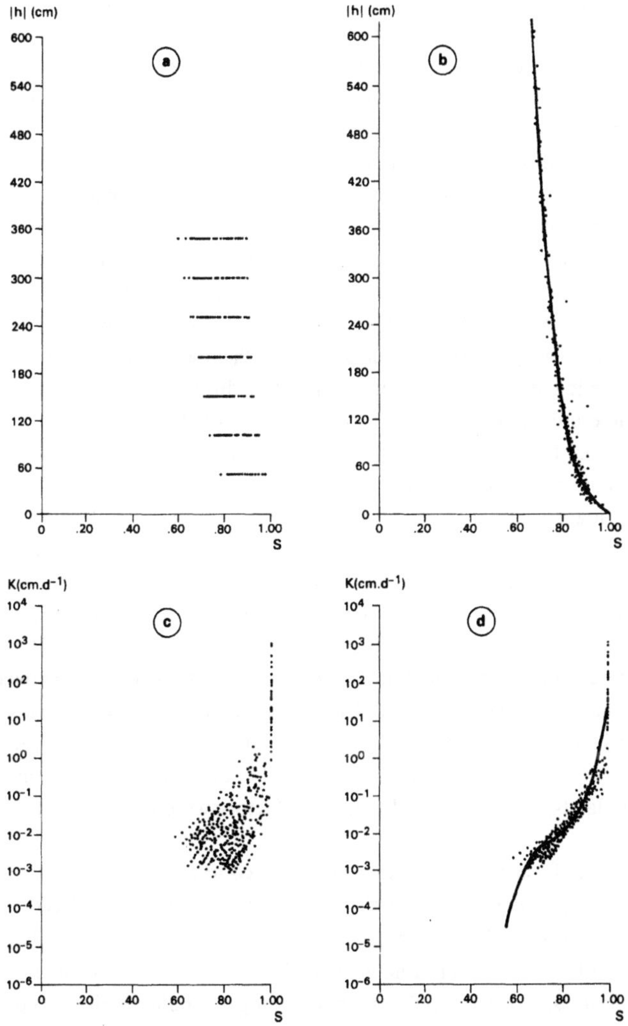

Figure 4. Unscaled (a) and scaled (b) water retention data and unscaled (c) and scaled (d) hydraulic conductivity data for the fine-textured soil group.

The fact that scaling is more effective in reducing variability of the moisture retention function as compared to the hydraulic conductivity function agrees with findings of Warrick et al. (1977b) and Hopmans (1987). This result is attributed to the fact that changes in the degree of saturation (S) have usually a greater effect on K than on h, as well as to the fact that measurement of h is less complicated and therefore more reliable than measurement of K.

The assumption that the three soil groups could be considered as similar media was tested by comparing the scale factors calculated from h data, i.e. $a_r(h)$, with those calculated from K data, i.e. $a_r(K)$, for all three soil groups together. Because individual soil groups are quite variable in soil texture (Table 1), and because no distinction is made between type of horizon it can not be expected that the two sets of scale factors follow exactly the 1:1 line. However, the two sets of scale factors were correlated significantly with R^2 equals 0.74 (R is the correlation coefficient). The slope of the regression line, emanating from the origin was calculated to be 1.02. These results agree with those reported by Warrick et al. (1977b), Simmons et al. (1979), Russo and Bresler (1980) and Hopmans (1987). Lack of correlation between the two sets of scale factors reported by Rao et al. (1983) is probably due to the fact that they used soils that were classified by the same taxonomic name, but consisted of distinctly different horizons, consisting of sand and clay in the same profile. Based on the results of this study, distribution functions of scale factors calculated from h data can be used to describe variability of water retention as well as variability of hydraulic conductivity.

Scale factors were lognormally distributed which agrees with findings of Nielsen et al. (1973), Warrick et al. (1977b), Russo and Bresler (1980) and Hopmans (1987). Table 3 presents the mean, variance and coefficient of variation of scale factors calculated from h data for each soil group. The coefficients of variation are within the range reported by Warrick et al. (1977b), Russo and Bresler (1980) and Hopmans (1987). Jury et al. (1987a) and Hopmans (1987) reported spatial correlations for both hydraulic conductivity and water retention over distances in the order of about 10 m. This implies that in the analysis of scale factors of nearby measurements spatial correlations of these scale factors should be taken into account (Jury et al., 1987b). However, because the hydraulic functions used in this study were measured at locations all over the Netherlands, it was assumed that scale factors were uncorrelated and statistically independent.

Distribution functions of scale factors (Table 3) were used to generate 100 new hydraulic functions for each soil group. The three functional criteria which were described earlier, were calculated for these new functions. This procedure is comparable to studies (e.g. Clapp et al. 1983) which use Monte Carlo simulations to calculate the distribution of soil moisture using scale factors as random variables. However, rather than assuming a certain distribution of scale factors the actually calculated distribution (Table 3) is used in this study. Functional criteria calculated on the basis of newly generated hydraulic functions, and on the basis of measured hydraulic functions are summarized in Table 4. Comparison among the three soil groups shows that all functional criteria are different, indicating differences in soil hydraulic functions of the three soil groups. Mean values of functional criteria using unscaled and scaled hydraulic function are fairly

Table 3. Mean (m_2), variance (sd_2^2) and coefficient of variation of scale factors calculated from moisture retention data.

soil group	mean	variance	coefficient of variation
coarse-tectured	1.00	0.56	0.75
medium-textured	1.02	0.37	0.60
fine-textured	1.04	1.07	0.99

Table 4. Mean (m), standard deviation (sd) and percentage variation (pv) of the functional criteria explained by the scaled hydraulic functions.

soil group	number of functions	travel time (d)			depth of water table (cm)			downward flux of water (cm d^{-1})		
		m	sd	pv	m	sd	pv	m	sd	pv
coarse-textured										
unscaled	105	118.7	62.3		146.4	52.3		38.6	163.6	
scaled	100	123.5	40.3	65	147.5	28.3	54	31.1	24.8	15
medium-textured										
unscaled	43	237.4	36.7		90.5	36.0		1.4	3.5	
scaled	100	234.3	33.6	92	88.9	6.5	18	0.9	1.5	43
fine-textured										
unscaled	49	339.5	66.9		44.7	25.3		1.0	10.7	
scaled	100	335.8	63.9	96	50.5	9.4	37	0.7	2.7	25

close. However, the functional criteria show smaller variations, as expressed by standard deviations, for scaled hydraulic functions as compared to unscaled hydraulic functions. This is especially true for Depth of water table (L) and Downward flux of water (F). Ranges in the percentage variation in functional criteria of unscaled hydraulic functions which would be explained by variation in functional criteria of scaled hydraulic functions are from 96% for Travel time (T) in the fine-textured soil group to 15% for Downward flux of water (F) in the coarse-textured soil group. The ranges in percentage explained variation for different functional criteria within the same soil group was due to the fact that effectiveness of scaling in reducing scatter of the original data differed for different h- and K values along the h(S)- and K(S) function line (Figures 2, 3 and 4). Since different functional criteria use different h(S)- and K(S) values, the percentage of explained variation differs.

Although scaling is clearly an attractive procedure to describe variability in hydraulic functions in the form of a single random variable, it underestimates at the same time the variation in functional criteria. Explained variation in functional criteria is maximal 100% only when scaling is 100% effective and scaled hydraulic functions would coincide completely with scaled mean hydraulic functions.

5. CONCLUSIONS

Scaling was successfully used to reduce variation in measured hydraulic functions into a narrow band around the scaled mean hydraulic functions of a large data base containing water retention and hydraulic conductivity data of soils in the Netherlands. In order to be successful, soils have to be grouped in sufficient large groups. Scaling proved to be an attractive technique to simplify the description of variability in soil hydraulic functions in the form of scaled mean hydraulic functions and distribution functions of scale factors. However, at the same time scaling underestimates existing variability. Results obtained for the three soil groups can be used to estimate the minimal variability in model output. Expansion of the data base with new experimental hydraulic functions will facilitate the distinction of different subgroups within the three soil groups distinguished in this study.

References

Ahuja, L.R., J.W. Naney and D.R. Nielsen, 1984: *Scaling soil water properties and infiltration modeling*. Soil Sci. Soc. Am. J., **48**, 970-973.

Arya, L.M.,D.A. Farrell and G.R. Blake, 1975: *A field study of soil water depletion patterns in presence of growing soybean roots. I. Determination of hydraulic properties of the soil.* Soil Sci. Soc. Am. Proc., **39**, 424-430.

Boels, D., J.B.H.M. Van Gils, G.J. Veerman and K.E. Wit, 1978: *Theory and system of automatic determination of soil moisture properties and unsaturated hydraulic conductivities*. Soil Sci., **126**, 191-199.

Bouma, J., 1982: *Measuring the hydraulic conductivity of soil horizons with continuous macropores*. Soil Sci. Soc. Am. J., **46**, 438-441.

Bouma, J., 1984: *Using soil morphology to develop measurement methods and simulation techniques for water movement in heavy clay soils*. p. 298-310. In J. Bouma and P.A.C. Raats (eds.), Proc. of the ISSS Sympos. on Water and solute movement in heavy clay soils. Wageningen, The Netherlands, 27-31 Aug. 1984. Publ. 37 Int. Inst. for Land Reclamation and Improvement (ILRI), Wageningen.

Bouma, J., C. Belmans, L.W. Dekker and W.J.M. Jeurissen, 1983: *Assessing the suitability of soils with macropores for subsurface liquis waste disposal*. J. Environ. Qual., **12**, 305-311.

Breeuwsma, A., J.H.M. Wösten, J.J. Vleeshouwer, A.M. van Slobbe and J. Bouma, 1986: *Derivation of land qualities to assess environmental problems from soil surveys*. Soil Sci. Soc. Am. J., **50**, 186-190.

Bresler, E. and G. Dagan, 1979: *Solute dispersion in unsaturated hetero-geneous soil at field scale: II. Applications*. Soil Sci. Soc. Am. J., **43**, 467-472.

Clapp, R.B., G.M. Hornberger and B.J. Cosby, 1983: *Estimating spatial variability in soil moisture with a simplified dynamic model*. Water Resour. Res., **19**, 739-745.

Dirksen, C., 1979: *Flux-controlled sorptivity measurements to determine soil hydraulic property functions*. Soil Sci. Soc. Am. J., **43**, 827-834.

FAO, 1985: *Guidelines: land evaluation for irrigated agriculture*. FAO Soils Bulletin, **55**, Rome.

Gardner, W.R., 1958: *Some steady state solutions of the unsaturated moisture flow equation with application to evaporation from a water-table*. Soil Sci., **85**, 228-232.

Hopmans, J.W., 1987: *A comparison of various methods to scale soil hydraulic properties*. J. of Hydrol., **93**, 241-256.

Hopmans, J.W. and J.N.M. Stricker, 1989: *Stochastic analysis of soil water regime in a watershed*. J. of Hydrol., **105**, 57-84.

Jury, W.A., D. Russo, G. Sposito and H. Elabd, 1987a: *The spatial variability of water and solute transport properties in unsaturated soil. I. Analysis of property variation and spatial structure with statistical models*. Hilgardia, **55**, 1-32.

Jury, W.A., D. Russo and G. Sposito, 1987b: *The spatial variability of water and solute transport properties in unsaturated soil. II. Scaling models of water transport*. Hilgardia, **55**, 33-56.

Miller, E.E. and R.D. Miller, 1955: *Theory of capillary flow: I. Practical implications*. Soil Sci. Soc. Am. Proc., **19**, 267-271.

Miller, E.E. and R.D. Miller, 1956: *Physical theory for capillary flow phenomena*. J. Appl. Phys., **27**, 324-332.

Montgomery, D.C. and F.A. Peck, 1982: *Introduction to linear regression analysis*. John Wiley & Sons, New York. pp. 504.

Nielsen, D.R., J.W. Biggar and K.T. Erh, 1973: *Spatial variability of field-measured soil-water properties*. Hilgardia, **42**, 215-259. (Plus separate appendices).

Rao, P.S.C., R.E. Jessup, A.C. Hornsby, D.K. Cassel and W.A. Pollans, 1983: *Scaling soil microhydrological properties of Lakeland and Konawa soils using similar media concepts*. Agric. Water Manage, **6**, 277-290.

Richards, L.A., 1965: *Physical condition of water in soil*. In C.A. Black et al. (ed.) Methods of soil analysis, Agronomy, **9**. 131-137.

Russo, D. and E. Bresler, 1980: *Scaling soil hydraulic properties of a heterogeneous field*. Soil Sci. Soc. Am. J., **44**, 681-684.

Sharma, M.L. and R.J. Luxmoore, 1979: *Soil spatial variability and its consequences on simulated water balance*. Water Resour. Res., **15**, 1567-1573.

Simmons, C.S., D.R. Nielsen and J.W. Biggar, 1979: *Scaling of field-measured soil-water properties*. Hilgardia, **47**, 74-154.

van der Zee, S.E.A.T.M. and W.H. van Riemsdijk, 1987: *Transport of reactive solute in spatially variable soil systems*. Water Resour. Res., **23**, 2059-2070.

van Genuchten, M.Th., 1978: *Calculating the unsaturated hydraulic conductivity with a new closed-form analytical model*. Research Rep. 78-WR-08. Water Resources Program, Princeton Univ., Princeton, NJ.

van Genuchten, M.Th., 1980: *A closed-form equation for predicting the hydraulic conductivity of unsaturated soils*. Soil Sci. Soc. Am. J., **44**, 892-898.

Warrick, A.W., G.J. Mullen and D.R. Nielsen., 1977a: *Predictions of the soil water flux based upon field-measured soil-water properties*. Soil Sci. Soc. Am. J., **41**, 14-19.

Warrick, A.W., G.J. Mullen and D.R. Nielsen, 1977b: *Scaling field measured soil hydraulic properties using a similar media concept*. Water Resour. Res., **13**, 355-362.

Wösten, J.H.M., J. Bouma and G.H. Stoffelsen, 1985: *The use of soil survey data for regional soil water simulation models*. Soil Sci. Soc. Am. J., **49**, 1238-1244.

Wösten, J.H.M., M.H. Bannink, J.J. De Gruijter and J. Bouma, 1986: *A procedure to indentify different groups of hydraulic-conductivity and moisture-retention curves for soil horizons*. J. Hydrol., **86**, 133-145.

Wösten, J.H.M., M.H. Bannink and J. Beuving, 1987: *Water retention and hydraulic conductivity characteristics of top- and subsoils in the Netherlands: The Staring series*. Report No. 1932, Soil Survey Institute, Wageningen, The Netherlands.

Wösten, J.H.M. and M.Th. Van Genuchten, 1988: *Using texture and other soil properties to predict the unsaturated soil hydraulic functions*. Soil Sci. Soc. Am. J., **52**, 1762-1770.

J.H.M. Wösten, The Winand Staring Centre for Integrated Land, Soil and Water Research, P.O. Box 125, 5700 AC Wageningen, The Netherlands

Field-Scale Water and
Solute Flux in Soils
Monte Verità

KRIGING VERSUS ALTERNATIVE INTERPOLATORS: ERRORS AND SENSITIVITY TO MODEL INPUTS

A.W. Warrick, R. Zhang, M.M. Moody, D.E. Myers

This study deals with interpolations through time and space. Two data sets are utilized, soil water content values at the 25 cm depth for a 45 hectare and an artificially-generated set of random but spatially correlated values. Interpolators include punctual kriging, a nonparametric estimator and co-kriging. Results show that although the model parameters depicting spatial interdependence (such as range and sill for a variogram) are highly dependent on the sample, the interpolations resulting from use of such models are not very sensitive to the model parameters. Co-kriging predictions were shown to be effective by using moisture content values for two different times as covariates. Additionally, an example illustrates the variogram for moisture content measured at one time can be adjusted and used for interpolations at a second time.

1. Objectives

Field-scale characterization of soils requires interpolations and extrapolations through time and space. Here we examine some alternative methods and compare performance based on a set of field-measured moisture contents and a set of simulated data. Questions addressed include:

a. How sensitive are the interpolator model parameters to the sample on which they are based? Included are considerations on numbers in a sample and spatial distribution.

b. How sensitive are estimated values to the model parameters?

c. How much help are auxiliary properties, such as are used by co-kriging or time invariance principles?

   Included in the analysis are the effects of sample number and locations. Also, the preservations of information from one time to another is examined as well as use of texture and water retention properties as supplementary information to predict water contents.

2. The Interpolators and the Data Sets

Interpolators chosen are kriging and a nonparametric regression model (NPR) (Watson, 1964; Yakowitz and Szidarovsky, 1985). For variogram models, parameters are determined in different ways, but for the main, are based on the negative log likelihood function and methodology outlined by Samper and Neuman (1989a):

$$\mathrm{NLL} = M \ln(2\pi) + \sum_{i=1}^{M} \ln\left(\sigma_i^2\right) + \sum_{i=1}^{M} \left(e_i / \sigma_i\right)^2 \tag{1}$$

with M the number of observations, e_i the difference between the observed and kriged values at point i based on the other M-1 observation and σ_i the kriging standard deviation at point i based on the other M-1 observations.

In order to minimize the NLL as given by Eq. 1, it is observed that

$$\begin{aligned} \mathrm{NLL} = M \ln(2\pi) + M \ln C + \sum_{i=1}^{M} \ln\left[\left(\sigma_{u,i}\right)^2\right] \\ + (1/C) \sum_{i-1}^{M} \left(e_i / \sigma_{u,i}\right)^2 \end{aligned} \tag{2}$$

where C is the sill and $\sigma_{u,i}{}^2$ is obtained from

$$\sigma_i^2 = C\, \sigma_{u,i}^2 \tag{3}$$

(cf. Warrick and Myers, 1987). Also, since e_i is independent of the sill C, then the minimum will occur when

$$C = (1/M) \sum_{i=1}^{M} \left(e_i / \sigma_{u,i}\right)^2 \tag{4}$$

For a zero nugget and C as in Eq. 4, the value of NLL in Eq. 2 depends only on the range of the variogram (or alternatively the integral scale). For simplicity, the effect of a nugget will not be considered, but the approach would remain the same if it were. Minimums of the NLL were found by using a golden section search to find the range and then the sill is evaluated from Eq. 4. The NPR estimator (Yakowitz and Szidarovsky, 1985) is given by

$$y_o = \frac{\sum_{i=1}^{M} y_i\, K\left(x_o - x_i,\, a\right)}{\sum_{i=1}^{M} K\left(x_o - x_i,\, a\right)} \tag{5}$$

where y_i and x_i refer to known values and locations, respectively and y_o and x_o are for the estimated value and location. The K is a kernel function of bandwidth "a" which we take here as an exponential function:

$$K(x, a) = (1/a) \exp(-x/a) \tag{6}$$

Other choices of the kernel could be made including a Gaussian distribution function. A discussion of the expected errors and variance of the error are given by Yakowitz and Szidarovsky (1985).

The field data was from the Maricopa Agricultural Center in central Arizona and consists of moisture content at 5, 10 and 25 cm at 45 sites on May 31 and 91 sites on June 12, 1988. Figure 1 shows the 45 sites which are on a 100 m regular grid as well as 46 additional sites chosen in two clusters to provide a more uniform distribution of couple distances of separation. Common sites for the two times were sampled with an Oakfield probe of 2.5 cm inner diameter. Supporting data for texture and moisture retention are also available but will not be used here. Figure 2 shows the sample variogram for both dates and a fitted spherical model with zero nugget and ranges of 650 m for May 31 and 550 m for June 12 with sills of 18 and 15.5, respectively. The variogram was based on the 45 data points. The artificial data set is that of Samper and Neuman (1989b) and consists of 200 "random" points on a 30 x 30 unit area generated with an exponential variogram with unit sill and range of 3.

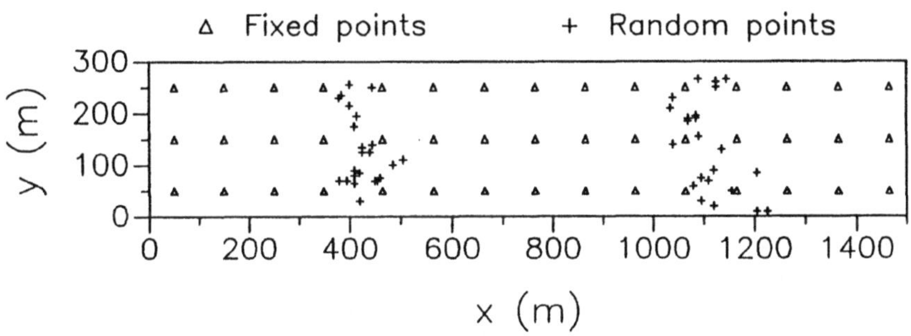

Figure 1: Sampling locations for moisture contents in the 45 hectare area.

3. Sensitivity of Model Parameters to the Specific Sample Chosen

The first example addresses the question of how the sample affects the model parameters of the interpolators. For this purpose, 30 subsamples were chosen:

10 for M = 15 points, 10 for M = 30 and 10 for M = 45 from the field moisture contents at 25 cm for June 12. Results are in Table I.

A spherical variogram model is used and Eq. 5 is used for the NPR kernel. Overall the estimates of the model parameters vary less from sample to sample as the sample size (M) increases. Both ranges and sills chosen by the maximum likelihood estimator in some cases are quite different from what would be estimated by a visual examination of Figure 2 although the tendency overall is consistent. For many of the subsamples, the range is shortened and the sill exaggerated, similar to the "Type I" error pointed out by Samper and Neuman (1989a, esp. Fig.20). The exercise was repeated for the synthetic data set for M of 25, 50 and 100 with similar results as shown in Table II. The fact that the NPR band width is different than the variogram range is not surprising, as they are selected on different principles.

4. Sensitivity of the Estimated Values to the Model Parameters

For the second example, the models given in Table II are used to calculate the root mean square errors (RMSE):

$$RMSE = \sqrt{(1/N) \sum_{i=1}^{N} (y_i^* - y_i)^2} \quad (7)$$

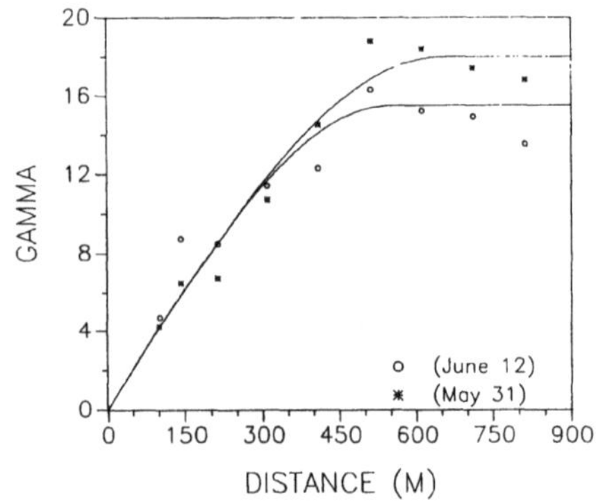

Figure 2: Variograms for May 31 and June 12 moisture contents.

where y_i^* and y_i are predicted values and measured values for points not used for the model determi tion, respectively. For all three values of M, the kriging values for the moisture contents (Table III) h a lower minimum and maximum value, although the upper values tend to be well within the range o other interpolators. The NPR values drop steadily as M gets larger, which is a desirable feature. small M, the inverse distance weighting (IDW) is as good as for the NPR, but for M = 45 it is not. I cases, the spatial estimators offer an improvement over the sample standard deviations (Table I).

Table I: Range of values for 10 repeated samples for the June 12 water content data.

M	mean	Sample s.d.	Variogram range	(Spherical) sill	NPR bandwidth
15	18.4-21.1	2.72-3.85	80-554	7.94-45.1	0.48-63.8
30	18.5-20.2	2.70-3.95	274-483	8.12-29.8	15.8-49.7
45	19.0-19.8	2.88-3.36	245-673	9.64-38.7	14.6-25.0

Table II: Range of values for 10 repeated samples for the simulated data of Samper and Neuman (1989a).

M	mean	Sample s.d.	Variogram range	(Exponential) sill	NPR bandwidth
25	0.96-1-31	0.73-1.23	0.71-6.85	0.52-1.60	1.50-14.5[1]
50	0.88-1.30	0.78-1.10	1.81-6.01	0.67-1.40	0.888-3.86
100	0.84-1.17	0.82-0.99	2.00-4.40	0.65-1.04	0.585-1.47

[1] Does not inclde 1 sample which did not converge.

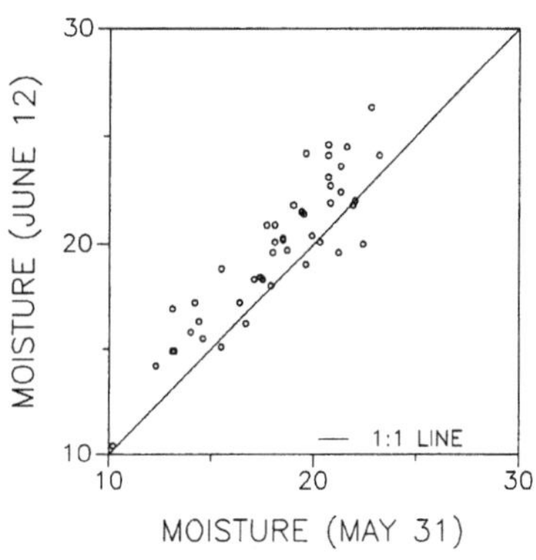

Fig. 3: Scatter diagram of soil moisture data for May 31 and June 12.

Results for the synthetic data set are in Table IV and are similar to those for moisture content results. As the original model is known for the synthetic data, the RMSE values were recalculated using an exponential model of unit sill and range of three. The results are given in Table IV and show that the RMSE is about the same as using the individually determined models.

5. Use of Auxiliary Properties

For the final comparison, 61 moisture contents for June 12 were estimated by using 30 values and 45 values from May 31 as a co-variate. Figure 2 shows the variogram models. The values tend to be time invariant (Vachaud, et al., 1985) with wet sites remaining wetter and drier staying drier for each of the two times. This is shown by the scatter diagram of Figure 3 which had a correlation r = 0.91 between the two times. Although the surface was considerably drier for May 31 than for June 12, the moisture contents at the 25 cm depth are close. One choice of using time-invariance is to simply use co-kriging, which greatly improves the results both for the RMSE and the kriging variance (see Table V).

Table III:Range of RMSE for the June 12 water content. In all cases N = 46 in Eq. 7.

M	Kriging	NPR	IDW[1]
15	1.88-3.12	2.13-3.54	2.13-3.33
30	1.79-2.36	1.92-2.81	2.03-2.77
45	1.74-2.21	1.87-2.31	2.04-2.60

[1] Inverse distance weighting with the same M known values.

Table IV: Range of RMSE for the simulated data. In all cases N = 100 in Eq 7.

M	Kriging	NPR	IDW[1]
25	0.784-0.998 (0783-0.977)[2]	0.822-1.02	0.771-1.00
50	0.700-0.898 (0.700-0.899)[2]	0.784-0.925	0.712-0.893
100	0.711-0.815 (0.692-0.811)[2]	0.719-0.828	0.746-0.850

[1] Inverse distance weighting with the same M known values.
[2] Using the original model.

As a final exercise, a variogram model for the June 12 data was obtained by multiplying the May 31 by the ratio of the sample variance. As shown in Table V (for the scaled variogram) the results are comparable to those for the original variogram, in fact, by chance the RMSE values are even lower.

Table V: Root mean square errors for June 12 moisture contents for predicting 61 points from 30 points

	RMSE	Kriging variance
kriging	2.92	13.6
co-kriging	1.81	8.6
inverse distance weighting	3.19	N/A
kriging (scaled variogram)	2.69	14.1

6. Summary and Conclusions

For the first objective, parameters for both the kriging and non-parametric models were found to be sensitive to the sample chosen. For the small samples (M=15 and 25), the range and sill varied by about an order of magnitude. The bandwidth varied by about two orders of magnitude (0.48-63.8) for the water content data. The range and bandwidth are less variable as M becomes larger and vary more typically by a factor of 1 to 2 or 1 to 3.

The kriging results for the RMSE were better in all cases as the sample size increased, although the accuracy was more influenced by the sample on which the interpolations are made than by the exact model. This is demonstrated most clearly for the synthetic results, for which the exact model gave results no better than the individually fitted models.

The final results illustrate two methods for interpolating spatially for different times. The first is by co-kriging with the moisture contents at the two times as co-variates. This is analogous to time-invariance methods although spatial interpolations can be included as well as site comparisons and rankings. The second method assumes the variograms are comparable and the sills adjusted with respect to the sample variance. Of course, the RMSE values are independent of the sills, but the scaling of the sills should lead to better overall performance.

A question often raised is whether kriging is worth the extra effort over more simple interpolators. These results show that RMSE values are only slightly better for kriging than for the other two interpolators. Similar results were found earlier (Warrick et al., 1988) for a larger variety of measured properties. The advantages of kriging exist primarily in the provision for additional error information, ability to include auxiliary information and the framework to predict from alternative sample support values. A final point to be made is the tendency towards "user friendly" software which allows easier and easier calculations.

References

Samper, F.J. and S.P. Neuman, 1989a: *Estimation of spatial covariance structures by adjoint state maximum likelihood cross validation 1. Theory*. Water Resourc.Res., **25**, 351-362.

Samper, F.J. and S.P. Neuman, 1989b: *Estimation of spatial covariance structures by adjoint state maximum likelihood cross validation 2. Synthetic experiments*. Water Resourc. Res., **25**, 363-371.

Vachaud, G., A. Passerat De Silans, P. Balabanis and M. Vauclin, 1985: *Temporal stability of spatially measured soil water probability density function*. Soil Science Soc. Amer. J., **49**, 822-828.

Warrick, A.W., and D.E. Myers, 1987: *Calculations of error variances with standardized variograms*. Soil Sci. Soc. Amer. J., **51**, 265-268.

Warrick, A.W., R. Zhang, M.K. El-Haris and D.E. Myers, 1988: *Direct comparison between kriging and other interpolators. Validation of low and transport models for the unsaturated zone, Ruidoso, NM:*

Watson, G.S., 1964: *Smooth regression analysis*. Sankhya Ser. A, **26**, 359-372.

Yakowitz, S.J. and F. Szidarovszky, 1985: *A comparison of kriging with non-parametric regression methods*. J. Multivariate Anal., **16**, 21-53.

A.W. Warrick, R. Zhang, Soil and Water Science, 429 Shantz #38, University of Arizona, Tucson, AZ 85721, USA.

M.M. Moody, D.E. Myers, Department of Mathematics, University of Arizona, Tucson, AZ 85721, USA.

Field-Scale Water and
Solute Flux in Soils
Monte Verità

SPATIAL AVERAGING OF SOLUTE AND WATER FLOWS IN SOIL

R. Webster and T.M. Addiscott

It is now generally recognized that hydraulic properties of the soil such as the conductivity and its reciprocal, the transfer time, are essentially random and spatially correlated variables. As such they are covered by the Theory of Regionalized Variables, which provides the framework for describing their statistical properties and the tools for estimation by kriging, decision-making and the design of sampling schemes. Distributions are notoriously skewed, and non-linear transformations are needed to stabilize variances, especially in the variogram, and for efficient estimation. Transforming to logarithms seems to work well. All the statistical analyses can be performed on them. However, the final estimates must usually be transformed back to the original scale for practical purposes. This is not straightforward, and the paper describes briefly the geostatistical procedure for obtaining unbiased estimates on the assumption that distributions are lognormal.

1. Introduction

The vertical fluxes of water and solute through the soil are usually measured on small supports, often cylindrical, with cross-sectional areas of 100 cm^2 to no more than about 1 m^2. The methods are reasonably refined and reliable, but they are expensive. Models designed to predict fluxes from more easily measured properties of the soil, whether by simulating processes within the soil or of the regression type, are also based on observations from small supports. Yet on the field scale we should like to know the values of fluxes at many sites or the average values over much larger areas. We wish to know what the vertical fluxes are through whole fields or larger regions.

It is unlikely that we shall ever know how the soil behaves physically between sampling supports over such a large area. So if we wish to infer values at unrecorded sites then we must resort to interpolation. Estimating the values over larger areas than those of the supports must also involve some kind of averaging.

Both interpolation and spatial averaging were traditionally aided by classification. The predictors, whether for individual points or for larger areas, were the class means. This was, of course, rather unsatisfactory. It assumed that the properties of interest were related to the classes of soil usually recognized on quite other characters. It ignored gradual change in the vicinity of mapped boundaries, and it paid no regard to spatial dependence within classes. Soil scientists, and especially soil physicists, are increasingly realizing that a more profitable approach to interpolation and estimation is embodied in the theory of regionalized variables (Matheron, 1965), specifically using kriging.

Regionalized variable theory treats variables that are distributed in one, two or three dimensions as continuous, random and spatially correlated. There may in addition be a deterministic component. At the scale of most studies of soil this last usually represents only a very small proportion of the total variation, and so in this paper we shall disregard it.

The theory provides a simple model to describe the variation in a random variable such as some characteristic of the soil. Denote the variable by Z and the particular value of it at $\mathbf{x}$, which represents the spatial coordinates of the site, by $z(\mathbf{x})$. Then the model of the variation is given by

$$z(\mathbf{x}) = \mu_v + \varepsilon(\mathbf{x}), \qquad (1)$$

where μ_v is the mean value of the property in a neighbourhood v and $\varepsilon(\mathbf{x})$ is a spatially correlated random variable with a mean of zero and variance defined by

$$\mathrm{var}[\varepsilon(\mathbf{x}) - \varepsilon(\mathbf{x}+\mathbf{h})] = \mathrm{E}[\{\varepsilon(\mathbf{x}) - \varepsilon(\mathbf{x}+\mathbf{h})\}^2], \qquad (2)$$

where $\mathbf{h}$ is a vector, the lag, separating two sites. The quantity $\gamma(\mathbf{h})$ is the semi-variance, and the function that relates γ to $\mathbf{h}$, the lag, is the variogram. The model leads directly to kriging for estimating values at unvisited sites with the same size and shape (support) as those on which the measurements were made, and over larger blocks. The former is a kind of interpolation, the latter yields mean values within blocks.

Kriging was developed mainly for mining, mineral prospecting and the like in which the aim was to estimate the concentrations and amounts of metal or ore in rock. It has proved equally profitable for predicting and mapping constituents of soil, and there are now many examples of its use.

Simple kriging is a form of weighted averaging in which the weights are chosen to avoid bias in the estimates, and subject to this condition to minimize the estimation variance. Thus when we estimate the value of Z at a place $\mathbf{x}_0$ to give $\hat{z}(\mathbf{x}_0)$ we want

$$\mathrm{E}[\hat{z}(\mathbf{x}_0) - z(\mathbf{x}_0)] = 0$$

and $\mathrm{E}[\{\hat{z}(\mathbf{x}_0) - z(\mathbf{x}_0)\}^2]$ to be a minimum.

In general if our variogram is accurate then we shall achieve this using the usual kriging formulation which can be found in the standard texts (e.g. Journel and Huijbregts, 1978; Webster and Oliver, 1990).

A method that produces unbiased estimations with minimum variance is attractive, and there is a danger that kriging will thereby be seen as solving all our problems by providing a unique best solution in any situation. This attitude is strengthened by the availability of certain "black boxes" for kriging. Given a particular situation — a set of data and a variogram and certain specific constraints — kriging will indeed give unique minimum variance estimates. In studies such as the one described by Warrick (1990) in this volume and those by Laslett et al. (1987) and by Voltz and Webster (1990) it has been found to perform better than other methods of interpolation in most instances.

Kriging, however, depends crucially on the variogram, and that is certainly not unique and often not well estimated. The results of kriging also depend on the assumptions we make about the distribution underlying the data and errors in them and any transformations we make.

2. The variogram

As above there is no unique variogram for a soil property. First, the variogram depends on the support of the measurements. In general the bigger the support the more variance each measurement encompasses and the less there is to appear in the variogram. Despite the search for a practicable support large enough to extend to the local sill or range of variation no one has yet demonstrated that he or she has achieved it. The variogram can also depend on the size of the region sampled. This may be fixed for a particular study, but if the study is extended then the variogram might embrace more variation.

A second feature of the variogram is that in practice we never know it exactly. We have first to estimate it, either as a cloud of pair-wise variances, or as a set of averages at specific lags. Then we must choose and fit a reasonable model to represent the observed variogram (see McBratney and Webster, 1986). This involves personal choice and sound judgement, aided by good numerical software such as MLP (Ross, 1987) and Genstat (Genstat 5 Committee,1987), both of which we use. It is not at all automatic.

3. Additivity

Most of the properties for which kriging is used are additive. For example, the concentration of salt in ten equal samples of soil when averaged would equal the concentration when the samples are bulked and thoroughly mixed. And the total amount of salt in the bulked soil would be 10 times the average amount in the ten samples.

When we are concerned with the transport characteristics of the soil, however, these simple relations do not hold. First, we cannot physically bulk the soil to obtain a combined value for a larger mass or volume: we shall want to combine our measurements in a way that applies to a larger undisturbed volume. Second, not all of the flow properties are additive either in a statistical sense or a physical one. So when averaging values we should be sure that the averages make physical sense.

For example, we may have measured the hydraulic conductivity, a rate, at several places. We can compute the arithmetic mean, and thereby estimate the mean conductivity for the region. This would be appropriate if we wished to estimate the amount of water passing through a layer of soil within the region in a given time under a fixed head. But suppose we wished to know the time required for a given amount of water to pass through the layer. The reciprocal of the mean conductivity is not the same as the mean of the transit times at individual places. And the actual transit times within a region are not physically independent — water will move laterally to the channel of least resistance as well as vertically.

4. Skew

The more serious problem soil physicists face with flow properties of the soil is the nature of their statistical distributions. Hydraulic conductivity, for example, is notoriously skewed, as is the pore size distribution. The soil of most regions contains sparse large cracks and channels with a more or less dense matrix between. The water flows preferentially in these channels. Computing arithmetic averages from data that are so strongly skewed inevitably means that a few very large values will affect the means disproportionately. Is this what we want?

If we want some general index of conductivity for a region then we might prefer the median to the mean. Several people have found that measured conductivity is approximately lognormally distributed. So we might take logarithms and compute their means. This is equivalent to taking geometric means.

If on the other hand we want a value that we can use to compute the yield of water from a region then we should average the conductivities themselves. The few large values will contribute most to an average, and this is just what we want.

From a statistical point of view, however, this is unsatisfactory because the arithmetic mean is an inefficient estimator when distributions are very skewed: the estimation error is very large. If the distribution is lognormal or approximately so then it is better to transform data to logarithms, compute the means and variances in the logarithms, and obtain an unbiased estimate of the mean by transforming back using the standard formula.

$$z = \exp\left(\bar{y} + \frac{1}{2}\hat{\sigma}_y^2\right), \qquad (3)$$

where $\bar{y}$ is the mean of the transformed values, i.e. $y_i = \ln z_i$ for all i, and $\hat{\sigma}_Y^2$ is the estimated variance of Y (see Aitchison and Brown, 1957).

This procedure assumes either that there is no error in the observations, or that the observational error is the same over the whole range of Z. If, however, the error varies in proportion to the value of Z then large values ought to carry less weight than small ones, and the geometric mean, $\bar{z} = \exp(y)$, might be closer to the true mean than that given by equation (3). Again there is a choice that depends on our prior knowledge of the situation.

5. Kriging estimators

When data are spatially dependent we have a further complication. The variogram computed from strongly skewed data is likely to be very crude: it is very sensitive to stray large values. Again it pays to convert to logarithms to stabilize the variances. The variogram is then computed from the transformed data, and values at unmeasured sites or over larger blocks are calculated by the usual kriging formulae. These will be unbiased estimates of the logarithms of the true values. As above, however, we may want unbiased estimates on the original scales, e.g. hydraulic conductivity. The back transformation analogous to equation (3), where there was no spatial dependence, is:

$$\hat{z}(\mathbf{x}_0) = \exp\left\{\hat{y}(\mathbf{x}_0) + \frac{1}{2}\hat{\sigma}^2_{\hat{y}}(\mathbf{x}_0)\right\}, \tag{4}$$

where $\hat{y}$ $(\mathbf{x}_0)$ is the estimated value of Y at $\mathbf{x}_0$ and $\overline{\sigma}^2_{\hat{y}}$ $(\mathbf{x}_0)$ is its estimation variance (Journel and Huijbregts, 1978). Unfortunately, as Journel and Huijbregts point out, this procedure can still lead to bias. To overcome this they multiply the back transformed value by a factor r. This factor is the ratio of the mean of the original data to the mean of the kriged values obtained using equation (4) within the region:

$$r = \frac{1}{n}\sum_{i=1}^{n} z(\mathbf{x}_i) \Big/ \frac{1}{k}\sum_{j-1}^{k} \hat{z}(\mathbf{x}_j). \tag{5}$$

6. Example

To give some feel for the problems we have taken results from a survey of hydraulic conductivity in the soil of one of our fields at Rothamsted. A 50 m x 50 m plot within the field was sampled using a nested design described by Oliver and Webster (1986) with pairs of points separated by lag distances of 0.5, 1.0, 2.0, 4.0, 8.0 and 16.0 m. There were 144 sampling points in total. At each point the saturated conductivity was measured as the distance fallen by a column of water in 1 minute after 1 hour of flow in holes 5 cm diameter.

Figure 1 shows a histogram of the resulting measurements. The frequency distribution is strongly skewed, and the curve that is fitted is that of the lognormal distribution. The fit is good. We can work in units of time to fall a given distance, and if we do then we obtain Figure 2. Since time is the reciprocal of rate this distribution is also approximately lognormal.

The components of variance were estimated for the spacings of the sampling design by analysis of variance, again as described by Oliver and Webster (1986), and the components accumulated to produce a variogram. The plotted points in Figures 3 and 4 are the result.

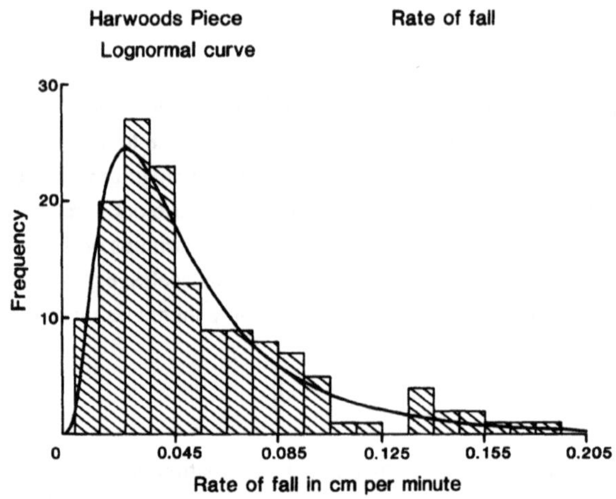

Figure 1: Histogram of rate of water flow.

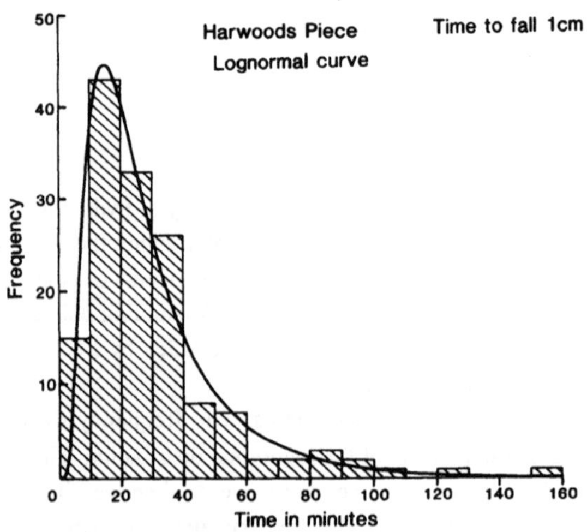

Figure 2: Histogram of transit time.

We have now to choose a model for the variogram and fit it, and the example illustrates the kind of problem that the analyst faces. If we work in the logarithms the solution appears straightforward, Figure 3a. The variogram looks to be linear and unbounded within the region sampled. Given that we transformed the data in the first instance to stabilize the variance, perhaps we should be satisfied with this solution. The best fitting model to the variogram computed on the original scale, however, is spherical (Figure 3b), which implies bounded variation within the region. Both cannot be correct. We could take the view that the plotted points are still essentially estimating a linear variogram. If we fit a linear model then we obtain the dashed line of Figure 3b, which is obviously relatively poor. In fact the residual sum of squares, one of the criteria for goodness of fit, is 0.000307 compared with 0.000202 for the spherical variogram. A third possibility is to treat the semi-variance at the largest lag as the least precise, or even ignore it because of its large leverage. If we then fit a linear model we obtain the solid straight line, which agrees closely with the spherical model over the first five estimates.

Figure 4 shows a further facet of the problem when we work in the reciprocals of the conductivity. The variogram of the logarithms is the same, of course. The variogram of the travel times, however, is more erratic than that of the conductivities, and no reasonable model can accommodate the semi-variance at 1 m. The best fitting model is now a power function, which is again unbounded, but of quite different form from the linear model. One would be loathe to disregard the semi-variance at the shortest lag because it is the most precisely estimated.

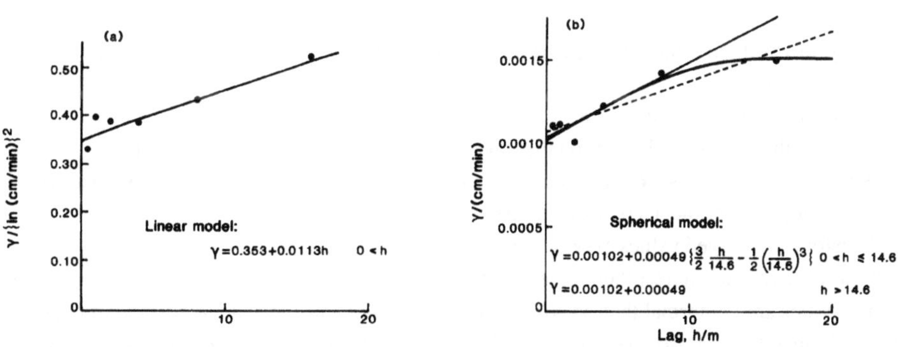

Figure 3: Variogram of rate of water flow on logarithmic scale (a) and on the original scale below (b). The curved line in (b) is that of the spherical model with the formula given. The dashed line is that of the best fitting linear model. The continuous straight line is the linear model fitted to the first five sample semi-variances only.

Clearly there is no clear cut model that can be taken to describe the spatial variation in this instance. The model that one fits will depend on one's prior knowledge of the general form of variation. If this is lacking then further sampling would be needed. In practice one would want more data from a more evenly distributed sample to estimate the variogram more precisely before going on to kriging. Fortunately, the precise form of the variogram does not seriously affect the kriging estimates. Most reasonable models will give similar results, though the estimation variances will differ more from one another.

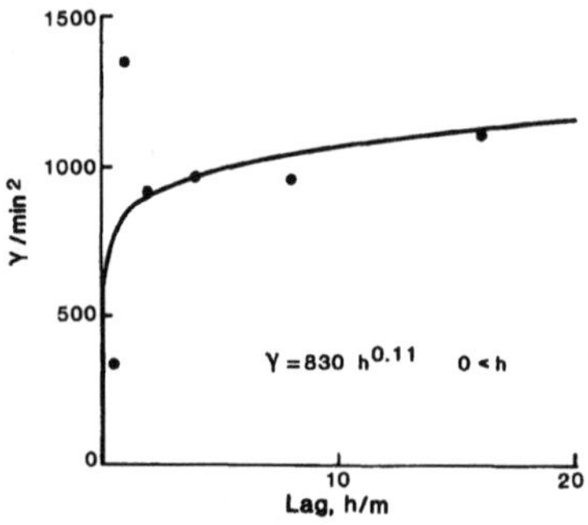

Figure 4: Variogram of transit time on original scale.

7. Conclusion

The way to estimate conductivities from spatially correlated measurements of conductivity seems to be as follows.

- Transform the data to logarithms.
- Compute the sample variogram on the transformed data and fit a suitable model.
- Estimate by kriging a set of values in the transformed scale for the region including values for particular points or blocks of specific interest.
- Transform the kriged values back to the original scale and correct for bias if necessary.

If transit times are of interest then these should be calculated from the conductivities first. If the conductivities are lognormal then so will be the transit times since they are reciprocal of conductivities. Transit times at isolated points can then be estimated by the same procedures as conductivities. However, estimating the transit time for blocks larger than the support of the sample may given incorrect results because of physical interaction.

Both the physics and statistics of water flux depend on conceptual models. They are of equal status. Both are our own interventions: they depend on our intellect as we try to interpret our observations, to build them into a coherent and self-consistent picture, and eventually to use them to predict. We shall never know whether they are right, but we must ever be ready to reject or modify them if they do not work.

Acknowledgements

We thank Dr A.P. Whitmore with whom we obtained the data to illustrate our paper and Mrs Joyce Munden for the graphics.

References

Aitchison, J. and Brown, J.A.C., 1957: *The lognormal distribution*. The University Press, Cambridge.

Genstat 5 Committee, 1987: *Genstat 5 reference manual,* Clarendon Press, Oxford.

Journel, A.G. and Huijbregts, C.J., 1978: *Mining geostatistics*. Academic Press, London.

Laslett, G.M., McBratney, A.B., Pahl, P.J. and Hutchinson, M.F., 1987: *Comparison of several spatial prediction methods for soil pH*. Journal of Soil Science, **38**, 325-341.

Matheron, G., 1965: *Les variables régionalisées et leur estimation*. Masson, Paris.

McBratney, A.B. and Webster, R. 1986: *Choosing functions for semi-variograms of soil properties and fitting them to sampling estimates*. Journal of Soil Science, **37**, 617-639.

Oliver, M.A. and Webster, R., 1986: *Combining nested and linear sampling for determining the scale and form of spatial variation of regionalised variables*. Geographical Analysis, **18**, 227–242.

Ross, G.J.S., 1987: *MLP Users' Manual*. Numerical Algorithms Group, Oxford.

Voltz, M. and Webster. R., 1990: *A comparison of kriging, cubic splines and classification for predicting soil properties from sample information*. Journal of Soil Science, **41**, in press.

Warrick, A., 1990: *Kriging versus alternative interpolators: errors and sensitivity to model inputs*. In: Proceedings of the Workshop on Field Scale Transport. Birkhäuser, Basel, in press.

Webster, R. and Oliver, M.A., 1990: *Statistical methods for soil and land resource survey*. Clarendon Press, Oxford.

R. Webster and T.M. Addiscott, Rothamsted Experimental Station, Harpenden, Hertfordshire AL5 2JQ, England.

Field-Scale Water and
Solute Flux in Soils
Monte Verità

CRITERIA FOR EVALUATING PESTICIDE LEACHING MODELS

K. Loague and R.E. Green

Mathematical models which permit query of "what if" scenarios are timely tools for assessing pesticide leaching potentials. Well-defined evaluation procedures for these important models have not, however, been firmly established. In this paper criteria for model evaluation are presented. Three examples of model testing are given. The models tested include a simple index of mobility and a deterministic-empirical simulation algorithm. The model testing examples make use of pesticide leaching data from Georgia and Hawaii.

1. INTRODUCTION

A pesticide leaching model is a good representation of reality only if it can be used to predict the observable phenomenon with acceptable accuracy and precision. Of course, no model can ever be detailed enough or excited with sufficient information to be totally valid for all pesticide leaching situations. Therefore, upon selecting what processes are to be modeled, a modeler must set a level of desired accuracy and precision for model validation. Standards for model performance are currently not well established. This paper is concerned with the demonstration of different criteria which can be used for evaluation of the match between observed and predicted pesticide leaching estimates.

The literature concerned with evaluation procedures for characterizing the performance of mathematical models in hydrology (e.g., Green and Stephenson, 1986; James and Burges, 1982; Martinec and Rango, 1989) and related fields (e.g., Tichelaar and Ruff, 1989; Willmott et al., 1985) is generally very useful for establishing evaluation protocols for pesticide leaching models. In this paper we focus on simple ranking procedures for qualitative evaluations and combinations of statistical criteria and graphical displays for quantitative evaluations. We use three examples to illustrate various model evaluation procedures and the associated nuances.

2. MODEL PERFORMANCE EVALUATION

There is virtually a continuum of mathematical models, in both scale and process detail, used to estimate pesticide leaching. Our objective here is to demonstrate model evaluation procedures which can be used to evaluate these different approaches. For example, the evaluation of an assessment from a pesticide leaching index, which does not include concentrations, can only be based on relative comparisons between the assessment and field observations. The evaluation of simulations from a physically-based pesticide leaching model, however, should include both statistical criteria and graphical displays. Assessment procedures which combine more than one measure of model performance are very useful for conducting comparative evaluations between competing models. In the remainder of this section alternative statistical criteria and graphical display techniques which can be employed in model performance evaluations are reviewed. The material covered here is taken, in part, from our earlier article (Loague and Green, 1990a).

Statistical Criteria

One statistical criterion for model performance evaluation is a comparison of summary statistics (mean and standard deviation) for observed and predicted summary variables gleaned from concentration profiles. Examples of summary variables for a leaching pesticide include (a) total mass, (b) center of mass, (c) peak concentration, (d) time for a critical concentration to leach to a depth of interest, and (e) depth of the leaching front. Summary variables should be selected based upon the problem at hand. A second statistical criteria for model evaluation is to use a test statistic to compare measured data against simulated results. A model's performance is judged acceptable if it is not possible to reject the hypothesis of no difference between observed and predicted values. Two types of error are possible using a test statistic at a given confidence level. Type I error is a risk to the model builder and corresponds to rejecting a true hypothesis. Type II error is a risk to the model user and corresponds to accepting a false hypothesis. Examples of test statistics include chi-square and Kolmogorov-Smirnov tests.

Analysis of residual errors also can be used to evaluate statistically the performance of pesticide leaching models by characterizing, for example, systematic under- or over-prediction. Several such measures are available; these include, for example: maximum error (ME), root mean square error (RMSE), coefficient of determination (CD), modeling efficiency (EF), and coefficient of residual mass (CRM). The mathematical expressions which describe these measures of analysis are given in Table I. The lower limit for ME, RMSE, and CD statistics is zero. The maximum value for EF is one. Both EF and CRM can become negative. CD is a measure of the proportion of the total variance of observed data explained by the predicted data. If EF is less than zero the model-predicted values are worse than simply using the observed mean. Several of the above statistics are sensitive to a few large errors especially in small data sets.

Graphical Displays
Statistical measures of model performance can have serious limitations. Graphical displays are often useful for showing trends, types of errors, and distribution patterns not identified with statistical measures. Several types of graphical display are possible. Four of the many possible methods of graphical display for the evaluation of pesticide leaching models are: (i) comparison of observed and predicted concentration profiles; (ii) comparison of ranges and medians of integrated values over prescribed depths for predicted and observed data; (iii) comparison of matched predicted and observed integrated values; and (iv) comparison of cumulative distribution functions for integrated values. Graphical technique (i) is used to judge the quality of model performance at specific sites. Graphical techniques (ii), (iii), and (iv) are used to evaluate model performance for several sites at once and are therefore not one-on-one tests. Systematic error, in the form of over- and under-prediction, can be gleaned easily from both (ii) and (iii). Spatial variations in observations and model prediction are represented by (iv).

3. EXAMPLES

In this section three examples of model testing are presented. The first example is for a simple pesticide mobility index. The second and third examples use the same dynamic conceptual pesticide leaching model. The material in the second example was, in part, previously reported (Loague and Green, 1990a).

I. Qualitative Evaluation: DBCP and EDB non-point leaching assessments for the near surface with the AF index; Pearl Harbor Basin, Oahu, Hawaii
Khan and Liang (1990) couple a simple chemical mobility index to a soil data base for Hawaii, using a geographic information system, to make "distributed" pesticide leaching assessments for the island of Oahu. There are at least two snags in the assessments reported by Khan and Liang, as pointed out by Loague and Green (1990b): (1) the mathematical model does not account for all known processes; and (2) the chemical, hydrologic, and soil data are sparse. In our effort, conducted at the same time as the work by Khan and Liang, the uncertainty in pesticide leaching assessments, due to uncertainties in input data, has been investigated and found to be significant (Loague et al., 1989a; Loague et al., 1990). In an ongoing effort, we are investigating the impact of model error on analyses of the type reported by Khan and Liang (Kleveno, 1990).
The index used by Khan and Liang (1989) is the attenuation factor (AF), proposed by Rao et al. (1985), given as

$$AF = \exp\left[\frac{-0.693 \cdot d \cdot RF \cdot \theta_{FC}}{q \cdot t_{1/2}}\right] \tag{1}$$

where d is the distance to groundwater or control depth from surface (L), RF is the retardation factor (dimensionless), θ_{FC} is the soil-water content at field capacity (volume fraction), q is the net groundwater recharge (LT^{-1}), and $t_{1/2}$ is the pesticide half-life (T). The retardation factor is given as

$$RF = 1 + \frac{\rho_b f_{oc} K_{oc}}{\theta_{FC}} + \frac{n_a K_h}{\theta_{FC}} \quad (2)$$

where ρ_b is the soil bulk density (ML^{-3}), foc is the soil organic carbon (mass fraction), K_{oc} is the organic carbon partition coefficient (L^3M^{-1}), n_a is the soil air-filled porosity (fraction, na = n - θ_{FC}), n is the soil porosity (fraction, n = 1 - ρ_b/ρ_p), ρ_p is the soil particle density (ML^{-3}), and K_h is the pesticide Henry's constant (dimensionless). The range of possible values for AF is between 0 and 1. For nonsorbed, nonvolatile pesticides RF = 1; with increasing K_{oc} and/or K_h, RF becomes larger. In general, the larger the AF value is for a pesticide the more likely it is that it will leach. Scales which have been used to divide the AF index into mobility ranges (e.g., Khan and Liang, 1989) are, to date, essentially arbitrary.

There is considerable interest (e.g. California, Florida, Hawaii) in using simple indices such as AF to screen and rank pesticides for regulatory purposes. If such techniques are to be used for decision making, they must be evaluated vis-a-vis ground truth whenever possible. In our example here we compare AF and RF estimates for DBCP and EDB for the Pearl Harbor Basin on Oahu with observations made in 1983 and 1985 (Tables II and III) and discuss some nuances of such comparisons. The input data for the AF and RF estimates are given in Table IV. The AF and RF estimates themselves are reported in Table V for the very near surface, corresponding to those estimates reported by Khan and Liang (1989). The standard deviations for the mobility estimates given in Table V are determined from first-order uncertainty analysis (Loague et al., 1989a; Loague et al., 1990).

Table VI summarizes the ranked mobility estimates for DBCP and EDB with the ranked field observations for 1983 and 1985 for the sites in the Pearl Harbor Basin. The observations (Table V) are ranked in order of the smallest concentrations. The mobility estimates for 1983 (Table II) and 1985 (Table III) are ranked in order of the smallest (RF) and largest (AF) estimates. Inspection of Tables V and VI leads to the following general comments:

(1) There is considerable variability and uncertainty in the RF and AF estimates.
(2) The comparisons are better for 1985 than for 1983.
(3) The comparisons are better for RF than for AF.
(4) The comparisons are better for EDB than for DBCP.
(5) The 1985 RF comparisons for EDB are perfectly aligned.

There are several problems in evaluating a pesticide leaching index as described here by field testing. First, and perhaps the most obvious, is the time span between a pesticide application and field sampling. If this duration is too short then comparing rankings of AF or RF estimates with measured concentrations, for various soil and land use classifications, will not be very useful for near-surface analyses. As shown in Table VII, there can be a strong relationship between pesticide residue and the time since the application. A second problem which complicates model evaluation is unexplained occurrences and concentration levels in the ground truth data base. In

Table II DBCP is observed at DM13 and DM60 even though the chemical was not previously applied. In Table VIII DBCP is observed at a higher concentration at D4201 in 1985 than in 1983 even though presumably it should be decreasing. A third problem is the use of near-surface mobility estimates to make assessments of potential groundwater contamination when the water table is at great depth (as in Hawaii). Due to a lack of hydrogeologic characterization, Khan and Liang (1989) concentrate only on the root zone in their leaching assessments. However, Khan and Liang do allude to the success of their system by noting the coincidence of surface areas mapped as very likely to leach EDB and the occurrence of EDB in the groundwater at the same general location. We are currently making AF estimates with depth for the deep holes listed in Tables II and III that will subsequently be compared with the observed data.

The evaluation procedure used here for non-quantitative leaching assessments, although not perfect, does permit a first-cut method for testing the usefulness of simple indices. This procedure does take into account the variability in soil and recharge information. As pesticide leaching assessments of the kind described by Khan and Liang (1989) become more common (and they will), their testing with field data becomes increasingly necessary.

II. Quantitative Evaluation: Atrazine simulations for the near surface with PRZM for multiple sites; Watkinsville, Georgia

For this example of model evaluation (as well as for the third example) pesticide mobility is simulated with the Pesticide Root Zone Model (PRZM). PRZM (Carsel et al., 1984) is a field-scale deterministic-empirical simulation tool for one-dimensional vertical transport of non-volatile agricultural chemicals through the unsaturated root zone. The soil-water movement algorithm employs a set of empirical drainage rules. The advection-dispersion equation is used to describe pesticide transport. The structure of PRZM is best suited to areas that are dominated by deep, well-drained, coarse-textured soils where the water table is near the surface. For a more complete description of PRZM the interested reader is directed to Carsel et al. (1984) and Loague et al. (1989b).

The observed data used in this example are from a 1.3-ha experimental catchment (P2) located in the Southern Piedmont near Watkinsville, Georgia. The P2 data base (Smith et al., 1978) includes atrazine residue measurements at several locations and depths on different dates after a controlled application of the pesticide. The application of PRZM to P2 generally satisfies the model assumptions and restrictions. The input data for the PRZM simulations are given by Loague and Green (1990a)..

The purpose of the model evaluations presented here is to illustrate alternative methods of model testing. Table IX summarizes the statistics (Table I) calculated for the integrated atrazine concentrations (observed and predicted data) from ten P2 sites above three specified depths on three separate dates after application of the pesticide. Figure 1 shows observed versus predicted atrazine concentrations at ten sites integrated to the specified depths 9, 25, and 91 days after application. Figure 2 shows cumulative distribution functions for observed and predicted atrazine concentrations using the same data as in Figure 1. The observed and predicted values in Figure 2 are normalized against the maximum observed value. Figure 3 is a set of bar graphs comparing the ranges of the observed and predicted atrazine concentrations used in Figures 1 and 2. The ME, RMSE, and EF statistics are best used with the graphical displays in Figure 1 while the CD and CRM statistics are best used with the graphical displays in Figures 2 and 3.

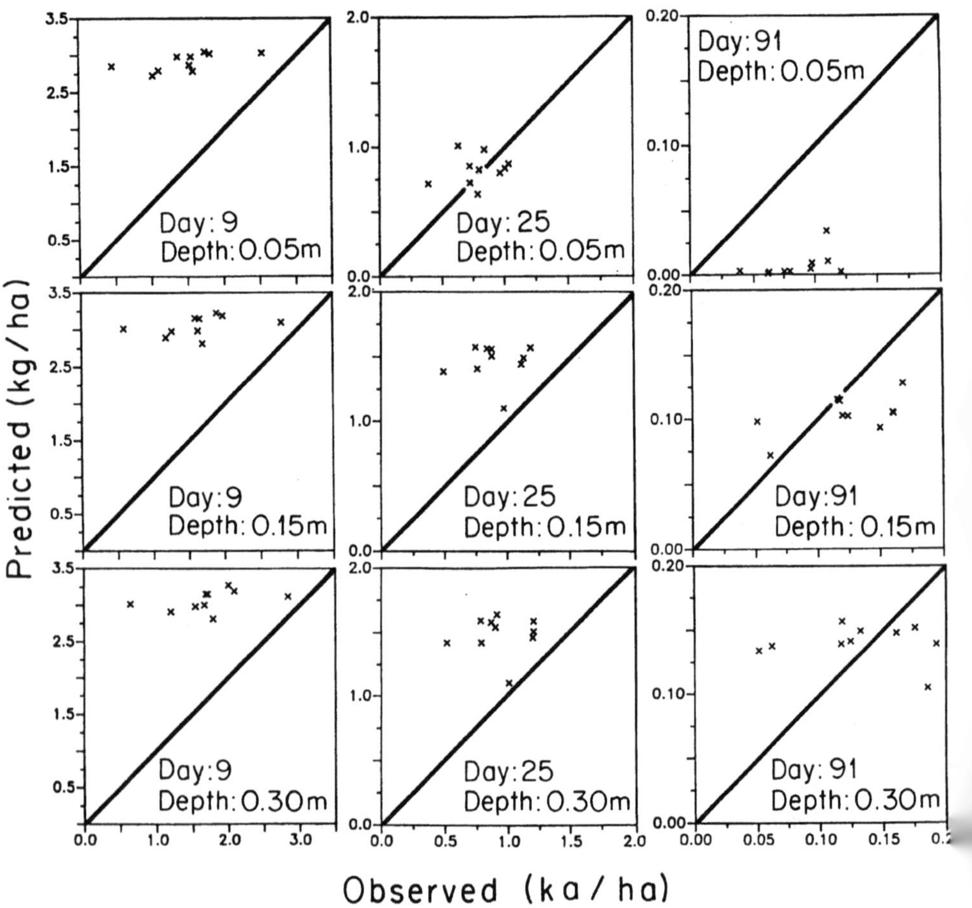

Figure 1: Observed versus predicted atrazine residues above given depths for all ten P2 sites (Loague and Green, 1990a).

Inspection of Table IX and Figures 1 through 3 leads to the following general comments:

1) The statistics are similar at each of the three depths on the same day. For example, on day 9 ME is equal to 2.39, 2.43, and 2.37 at depths of 0.05, 0.15, and 0.30 m, respectively. This observation holds best for day 9.
2) The statistics appear better at the later dates in some cases. For example, on days 9, 25, and 91 ME is equal to 2.39, 0.38, and 0.04, respectively, at 0.05 m. The concentrations have decreased, however, and the ME improvements are not real.
3) The model overpredicts early (day 9) and underpredicts later (day 91). This observation can be seen by reviewing the CD values. Figures 1 through 3 further illustrate how the model overpredicts at every site on day 9. Figure 2 also shows that the observed spatial variations between sites are not well predicted with the model. The CD values at 0.05 m oscillate from less than 1.0 on day 9 to greater than 1.0 on day 25 to less than 1.0 again on day 91, illustrating the transient nature of model performance.
4) The EF values indicate an overall poor model performance. Only one EF value is greater (slightly) than 0.0.
5) The model performance on day 25 at 0.05 m appears to be either quite good (Figures 1-3) or quite bad (EF). This example illustrates the need to combine statistical criteria with graphical displays in model evaluation.

The performance of the uncalibrated PRZM, as reported here for P2, is generally quite poor. Of course, it would be possible to calibrate the model for the P2 leaching events, using either selected sites or early times, and then attempt to validate the model at the remaining sites or later times. It is our opinion that the performance of PRZM could be improved for the P2 sites by the curve fitting exercise of parameter estimation. The suggested improvement for PRZM performance with calibration, however, will be complicated by input errors. Even though the P2 data base is an excellent one, there are limitations for using it to validate a model. Even with perfect field data it will not be possible to validate PRZM exactly due to the simplifying assumptions upon which it is based.

III. Quantitative Evaluation: Simulations of deep penetration of EDB with PRZM for a single site; Pearl Harbor Basin, Oahu, Hawaii

In a recent study we (Loague et al., 1989b; Loague et al., 1989c) conducted an initial qualitative evaluation of PRZM using the deep EDB (1983 and 1985) profiles shown in Figure 4 from the D4201 hole identified in Table II. In this work we applied PRZM beyond its intended (near surface) range to determine whether it could provide reasonable estimates of peak concentrations for leaching pesticides in highly structured soil and fractured rock. Here we quantitatively evaluate the performance of PRZM for one of our past simulations (see Figure 5) and for a series of 75 new calibration simulations. Because EDB is a volatile chemical, we preprocess short-term volatilization effects in the simulations reported here with a separate model as described by Loague et al. (1989b).

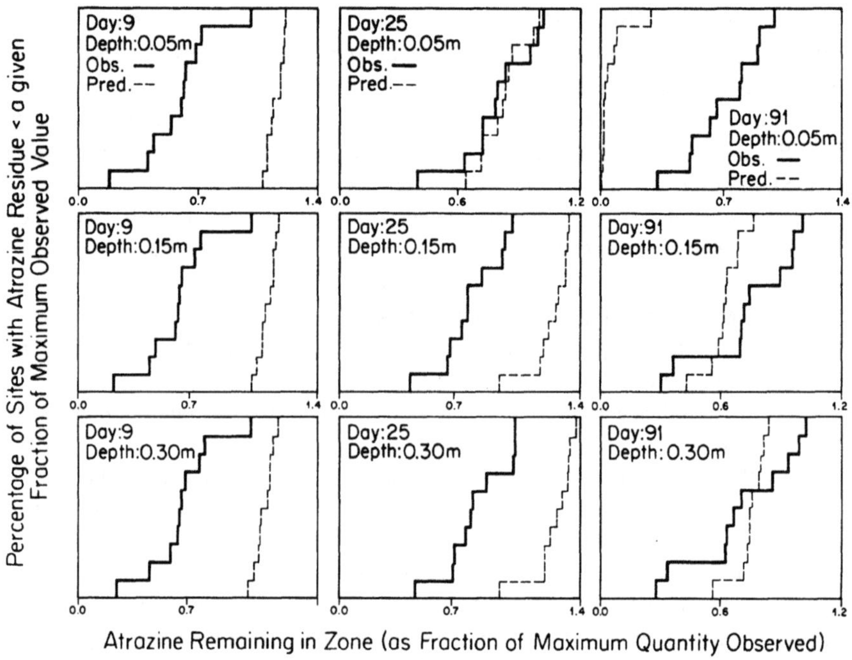

Figure 2: Cumulative distribution functions for observed and predicted atrazine residues above given depths for all ten P2 sites (Loague and Green, 1990a).

The calibration simulations consist of trial-and-error adjustments to two parameters: (i) the pesticide decay rate, and (ii) the hydrodynamic dispersion coefficient. All other parameters used by PRZM are as given for Case H in Loague et al. (1989c) and Case A in Loague et al. (1989b). The modeling strategy we follow here is to calibrate PRZM against the 1983 EDB profile and then test the validity of the calibration using the 1985 profile. Table X summarizes predicted EDB concentration profile characteristics (two summary variables) and statistical evaluations for the 1983 and 1985 simulations. The statistics presented here, in contrast to those in the previous example for atrazine at several locations, are for EDB profiles from a single site. The simulations are presented in ranked order relative to concentration profile characteristics and statistical results in Tables XI and XII, respectively. The average rankings given in Tables XI and XII are the ranked simple averages for the ranked individual profile characteristics and statistical results. For example, in Tables XI and XII, the 1983 and 1985 averages are averaged and then ranked to yield the total averages.

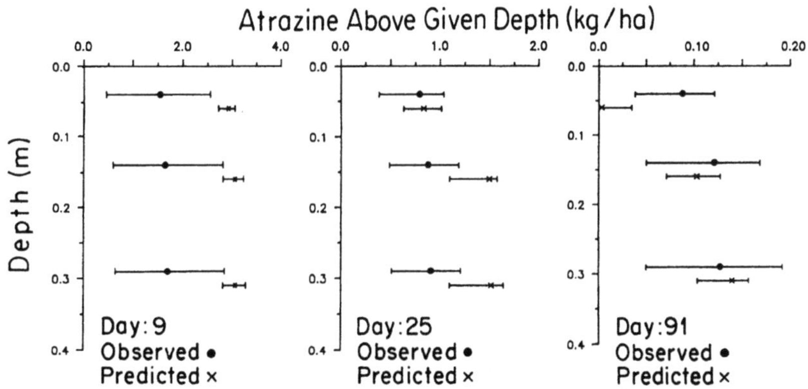

Figure 3: Bar graphs for observed and predicted atrazine residues above given depths for all ten P2 sites showing maximum, minimum, and median values (Loague and Green, 1990a).

Perusal of the 1983 results in Tables XI and XII indicates that for the various evaluation criteria the resulting best calibration simulations are different. Perusal of the 1985 results in Tables XI and XII shows that the model performance did not match the 1983 performance and that the best simulations for each evaluation criteria were not the same as for 1983. A closer inspection of Tables XI and XII leads to a few more comments about the calibration simulations:

1) The original simulation (Figure 5) was not evaluated as the best with any of the criteria.
2) Based upon the depth-to-peak concentration criterion, the best simulations (#'s 16-18, 45-46, 56-58, 68-71) have larger hydrodynamic dispersion coefficient values and smaller decay rates than the original case. There is obviously a trade-off between the two parameters.
3) Based upon the peak concentration criterion, the best simulation (#19) has a hydrodynamic dispersion coefficient more than twice that in the original simulation. Simulation number 19 also looks good based on the depth-to-peak criterion (average rank = 2).
4) Based upon the five statistical criteria, the best simulations (#'s 66, 45, 58, 60, and 45) have larger hydrodynamic dispersion coefficient values than the original case (95, 55, 50, 85, and 55 cm^2/day for ME, RMSE, CRM, CD, and EF, respectively). Simulations 45 and 58 were also ranked first for the depth-to-peak criterion.
5) The decay rate varies from 0.0044 to 0.0035 for the best simulations evaluated with all criteria. It appears that for the range of values tested, the performance of PRZM, in this situation, is less sensitive to adjustments in the decay rate than in the hydrodynamic dispersion coefficient, although both are obviously important.

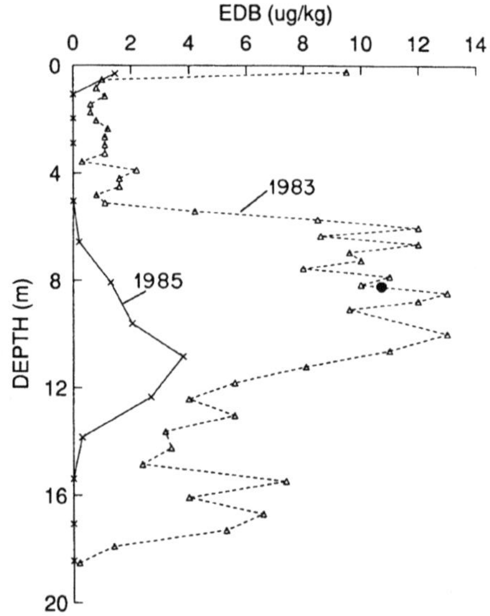

Figure 4: Observed EDB concentration profiles for D4201. The single large point represents the value taken as the observed peak concentration for 1983.

The interpretations of the 1985 simulations are generally very similar to those described above for 1983, although the model performance was much worse. The conflicting results for the alternative model performance measures shown here illustrate the need for multi-criteria evaluation procedures. It is interesting to note that based upon the average ranks shown in Table XII that simulation number 75 is the second best simulation for 1985 and the worst simulation for 1983. The best ranked (average) simulation (statistical criteria) for 1983 is number 56 which is the fifty-fifth best simulation for 1985. One can easily deduce from Tables X-XII and Figures 6-9 that the 1985 model validation, based upon the 1983 calibration, is quite poor.

The 1983 and 1985 EDB profiles can both be used together to calibrate PRZM. In Tables XI and XII the total averages are based upon ranking the combined averages from the 1983 and 1985 evaluations. For the statistical criteria (total averages, Table XII) simulations 18 and 19 are the best. For the concentration profile characteristic criteria (total averages, Table XI) simulations 19 and 20 are the best. The best simulations for 1983 based on individual criteria using profile characteristics and statistical measures are shown in Figures 6 and 7, respectively. The best simulations for 1983 and 1985, based on average criteria using profile characteristics and statistical measures, are shown in Figures 8 and 9, respectively.

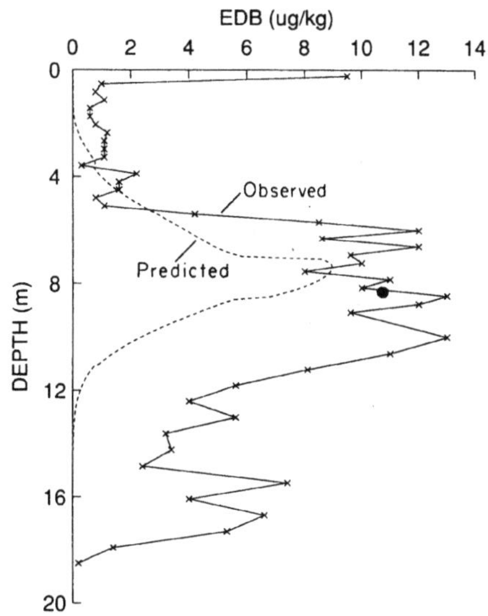

Figure 5: Observed vs. predicted EDB concentration profiles for 1983. The predicted profile is Case H from Loague et al. (1989c).

4. CONCLUSIONS

Pesticide leaching assessments which incorporate the results of mathematical models will increasingly be relied upon in the future for chemical licensing and regulation decisions. If these models are to be useful tools in the environmental management arena they will require rigorous testing with field data. In this paper we illustrate different facets of model evaluation for pesticide leaching models using three examples. In the first example we test a mobility index by comparing observed concentrations with relative mobility predictions for soils from the same

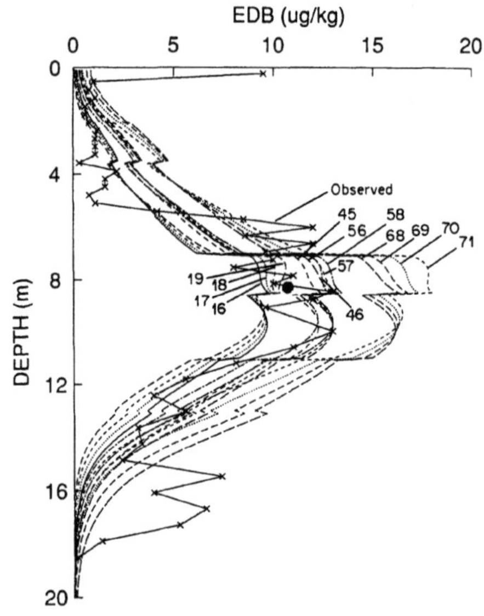

Figure 6: Observed vs. predicted EDB concentration profiles for 1983. The predicted profiles are those that ranked the best based on individual profile characteristics (Table XI).

taxonomic category. In the second example we test a simulation model using observed and predicted concentration data integrated over a suite of sites for a series of depths. In the third example we test a simulation model using observed and predicted concentration profiles from a single site. We have not recommended specific performance levels for qualitative or quantitative evaluations of pesticide leaching models. The intended use of an assessment or prediction must be considered when determining how stringent standards for model acceptance should be. Spatial and temporal variability in parameter uncertainties will impact upon the levels of model performance which can be expected. A given model deemed acceptable at a specific spatial scale may not be considered appropriate at other scales. Characterizing an acceptable model performance level is obviously dependent upon a number of interrelated factors.

The model performance results from any one statistical criteria or graphical display are not sufficient to pronounce a model's validation even if established levels of confidence were available.

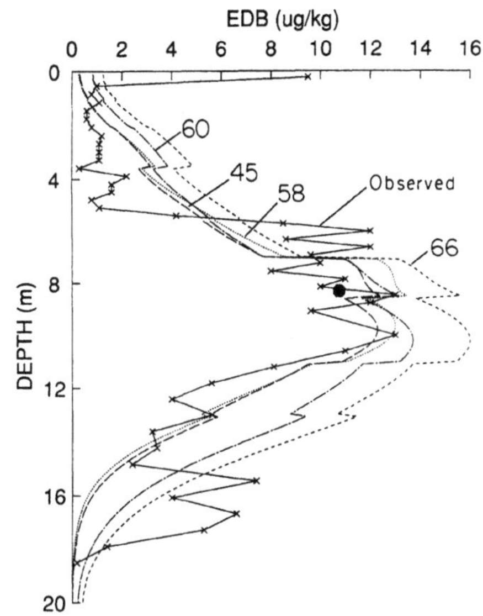

Figure 7: Observed vs. predicted EDB concentration profiles for 1983. The predicted profiles are those that ranked the best based on individual statistical criteria (Table XII).

Taken independently each method of evaluation can be limited by stringent assumptions. If even one assumption is violated, then sole reliance on the method is suspect. A model should be considered validated only if a set of performance tests are met. In this study we combined the results of several tests to select the best simulation in a trial and error calibration by ranking the results. Other examples of multi-criteria tests are given by Gilmour (1973) and Martinec and Rango (1989).

Generally, only a few of the parameters used in a dynamic simulation model such as PRZM are obtained directly from measurements at the locations where assessments of pesticide leaching are of interest. The missing information usually is estimated during the calibration of the model. Based on the initial results reported here, it appears that a single concentration profile from a single site is insufficient for a reliable model calibration and that several sites with profiles at various time intervals are needed. Chemical transformations complicate the model calibration and validation procedure tremendously. In fact, it may not be possible to validate certain pesticide leaching models for particular problems.

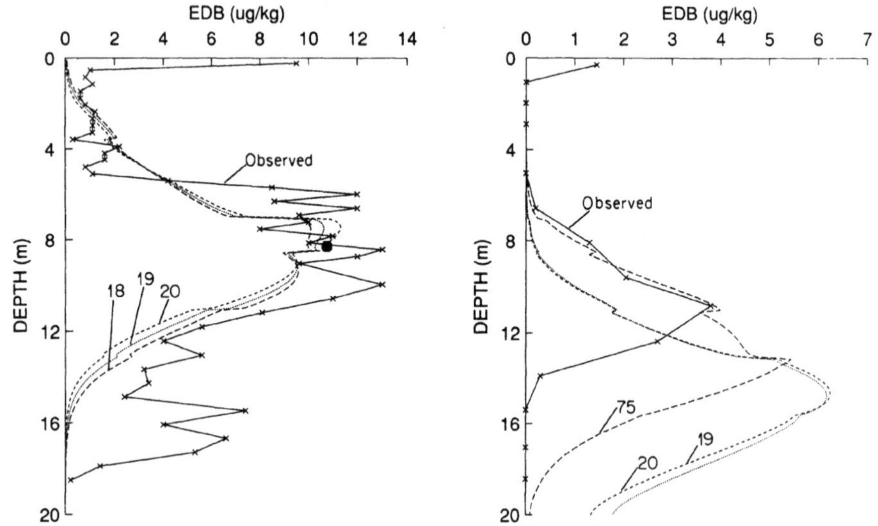

Figure 8: Observed vs. predicted EDB concentration profiles. The predicted profiles are those that ranked the best based on averaged profile characteristics (Table XI). (a) 1983, #18; (b) 1985, #75. Simulations 19 and 20 are the best for 1983 and 1985 evaluated together.

Improved field data sets are needed for model validation and comparison studies. There was generally more information for the model evaluation examples discussed here than is typical. Still each data set was far from perfect and often contained inadequate information. Before specific model testing protocols can be established, we must first identify what information should be collected. We should not collect information only because it has been collected in the past. For each problem we must decide what should be measured, the number of measurements to make, where the selected measurements should be made, and how often. Of course, each data collection program and model validation problem is separate and ultimately a question of the utility for the information and a finite budget.

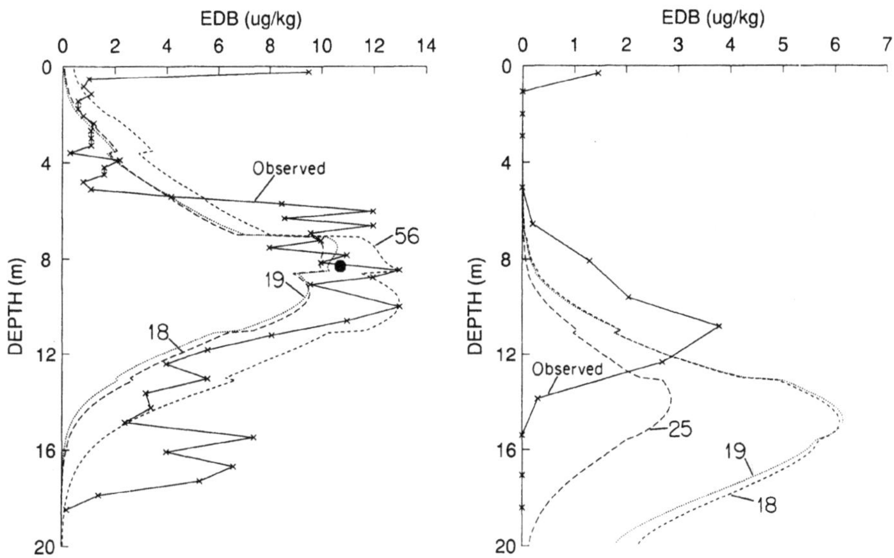

Figure 9: Observed vs. predicted EDB concentration profiles. The predicted profiles are those that ranked the best based on averaged statistical criteria (Table XII). (a) 1983, #56; (b) 1985, #25. Simulations 18 and 19 are the best for 1983 and 1985 evaluated together.

Model evaluation can be as important (if not more) as model selection. A model should be assumed suspect until it is proven correct. With regard to the evaluation of pesticide leaching models, it seems that there are three generally overlooked questions that need to be comprehensively addressed: (i) what has been done?, (ii) what should be done?, and (iii) what is likely to be done? We believe it would be useful if a standardized model evaluation protocol is adopted. Once established, such procedures may be used to catalog model performance for different models and applications.

Acknowledgments

The work reported here was supported in part by the Hawaii State Department of Agriculture. We are grateful to Fred Cheng, Tony Liang, and Rita Rausch for their assistance. The manuscript was typed by Susan Durham and Joan Van Horn.

Table I: Measures for analysis of residual errors.

Maximum error

$$ME = \mathrm{Max}\left|P_i - O_i\right|_{i=1}^{n}$$

Root mean square error

$$RMSE = \sqrt{\frac{1}{n}\sum_{i=1}^{n}(P_i - O_i)^2} \cdot \frac{100}{\overline{O}}$$

Coefficient of residual mass

$$CRM = \frac{\sum_{i=1}^{n} O_i - \sum_{i=1}^{n} P_i}{\sum_{i=1}^{n} O_i}$$

Coefficient of determination

$$CD = \frac{\sum_{i=1}^{n}(O_i - \overline{O})^2}{\sum_{i=1}^{n}(P_i - \overline{O})^2}$$

Modeling efficiency

$$EF = \frac{\sum_{i=1}^{n}(O_i - \overline{O})^2 - \sum_{i=1}^{n}(P_i - O_i)^2}{\sum_{i=1}^{n}(O_i - \overline{O})^2}$$

P_i = predicted values; O_i = observed values; $\overline{O}$ = mean of the observed data; n = number of samples.

Table II. Summary of the pesticide data base resulting from the 1983 drilling program in the Pearl Harbor Basin, Oahu, Hawaii (abstracted, in part, from Wong, 1983, 1987).

Site	Number of Holes	Average Hole Depth (m)	Soil Order	Elevation (m)	Average Annual Rainfall (mm)	DBCP Concentration (µg/kg)	DBCP Months after Application	DBCP Number of Previous Applications	EDB Concentration (µg/kg)	EDB Months after Application	EDB Number of Previous Applications
DM2	10	8.6	Oxisols	233	1000	0	na	-	2059	2	9
DM13	2	18.3	Oxisols	253	1000	2	na	-	10	18	8
DM60	3	18.6	Inceptisols	303	1100	1	na	-	4	46	5
D4107*	2	9.0	Oxisols	152	1000	8	104	2	1	63	0
D4111*	3	13.7	Oxisols	145	800	17	96	2	127	4	1
D4201	4	30.4	Ultisols	310	2000	13	84	4	8	27	0
D4213	3	18.6	Oxisols	240	1300	7	73	4	9	8	0

* Drip irrigated
na = never applied
All concentrations reported here sampled at 0.2 m.

Table III. Summary of the pesticide data base resulting from the 1985 drilling program in the Pearl Harbor Basin, Oahu, Hawaii (abstracted, in part, from Peterson et al., 1985).

Site	Number of Holes	Average Hole Depth (m)	Soil Order	Elevation (m)	Average Annual Rainfall (mm)	DBCP			EDB		
						Concentration (µg/kg)	Months after Application	Number of Previous Applications	Concentration (µg/kg)	Months after Application	Number of Previous Applications
DM8*	1	29.0	Inceptisols	275	1000	~0	na	-	10	24	5
DM31	1	21.4	Inceptisols	270	950	5	na	-	**	65	4
D4101*	1	18.3	Oxisols	165	900	~0	204	2	~0	27	1
D4111*	1	12.2	Oxisols	145	800	~0	123	2	~0	130	1
D4201	2	30.5	Ultisols	310	2000	18	108	4	2	32	0
D4213	1	30.5	Oxisols	240	1300	~0	97	4	~0	33	0

* Drip irrigated
**No data
na = never applied
Concentrations reported here were sampled at 0.3 m for D4101, D4111, and D4213; 0.4 m for D4201; and 0.6 m for DM8 and DM31.

Table IV: (a) Soil parameters* and recharge rates (for the Pearl Harbor Basin, Oahu, Hawaii) based on soil orders and land use. (b) Chemical properties used for pesticide mobility estimates with the AF and RF indices (Rao et al., 1985).

(a)

Soil Order	f_{oc} (dim)	θ_{FC} (dim)	ρ_b (kg/m^3)	ρ_p (kg/m^3)	q (m/day) Unirrigated Pineapple	Drip Irrigated Pineapple
Inceptisols	0.09	0.41	688	2727	0.0016	0.0023
Mollisols	0.02	0.37	1188	2862	0.0010	0.0015
Oxisols	0.03	0.43	1107	3005	0.0022	0.0030
Ultisols	0.04	0.40	1186	3170	0.0057	0.0063
Vertisols	0.02	0.42	1281	2934	0.0010	0.0016

* d = 0.5 m

(b)

Chemical Parameter	DBCP	EDB
K_{oc} (m^3/kg)	0.070	0.044
K_h (dim)	0.017	0.035
$t_{1/2}$ (days)	180	3650**

** This is not the value used for the PRZM simulations of EDB in example III.

Table V: RF and AF values and standard deviations* for DBCP and EDB in the Pearl Harbor Basin based on soil orders and land use.

	DBCP			EDB		
		AF			AF	
Soil Order	RF	Unirrigated Pineapple	Drip Irrigated Pineapple	RF	Unirrigated Pineapple	Drip Irrigated Pineapple
Inceptisols	11.7	0.003	0.017	7.7	0.83	0.88
	(7.4)	(0.011)	(0.047)	(4.6)	(0.10)	(0.07)
Mollisols	5.8	0.014	0.064	4.0	0.86	0.91
	(2.4)	(0.027)	(0.078)	(1.5)	(0.05)	(0.03)
Oxisols	5.6	0.120	0.210	3.9	0.93	0.95
	(2.7)	(0.130)	(0.165)	(1.7)	(0.03)	(0.02)
Ultisols	8.9	0.300	0.340	6.0	0.96	0.96
	(2.6)	(0.110)	(0.116)	(1.6)	(0.01)	(0.01)
Vertisols	4.2	0.032	0.120	3.0	0.89	0.93
	(1.8)	(0.049)	(0.110)	(1.1)	(0.04)	(0.03)

* Standard deviations are in brackets. The uncertainties in the RF and AF estimates are based only on the uncertainty in the soil and recharge data

Table VI. Comparison of ranked relative mobility estimates for DBCP and EDB with ranked field observations.

		DBCP			EDB		
Site	Year	Observed[1]	RF[2]	AF[3]	Observed	RF	AF
DM2	1983	1	1	3	7	1	3
DM13	1983	3	1	3	5	1	3
DM60	1983	2	3	4	2	3	4
D4107	1983	5	1	2	1	1	2
D4111	1983	7	1	2	6	1	2
D4201	1983	6	2	1	3	2	1
D4213	1983	4	1	3	4	1	3
DM8	1985	1	3	4	3	3	4
DM31	1985	2	3	5		3	5
D4101	1985	1	1	2	1	1	2
D4111	1985	1	1	2	1	1	2
D4201	1985	3	2	1	2	2	1
D4213	1985	1	1	3	1	1	3

[1] Ranked in order of the smallest observed concentrations (see Tables II and III).

[2] Ranked in order of the smallest estimates (see Table V). There are three soil orders and therefore three classes for both the 1983 and 1985 estimates.

[3] Ranked in order of the largest estimates (see Table V). There are four and five classes for the 1983 and 1985 estimates, respectively, based upon the different soil orders and recharge rates.

Table VII. Relationship between EDB concentration measured at the surface and time since application (abstracted from Table II).

EDB (μg/kg)	Months after Application
2059	2
127	4
27	8
10	18
8	27
4	46
1	63

Table VIII. Relationship between 1983 and 1985 DBCP and EDB concentrations measured at three sites in the Pearl Harbor Basin, Oahu, Hawaii (abstracted from Tables II and III).

	DBCP (μg/kg)		EDB (μg/kg)	
Site	1983	1985	1983	1985
D4111	15	~0	97	~0
D4201	13	18	8	2
D4213	7	~0	9	~0

Table IX. Model performance statistics for atrazine residues predicted with PRZM above given depths for ten P2 sites on three dates (after Loague and Green, 1990a).

Days after Application	Depth (m)	Statistic*				
		ME (kg/ha)	RMSE (%)	CRM	CD	EF
9	0.05	2.39	103	-0.98	0.13	-7.36
	0.15	2.43	94	-0.88	0.15	-6.68
	0.30	2.37	84	-0.79	0.16	-6.04
25	0.05	0.38	25	-0.05	2.48	-0.18
	0.15	0.89	68	-0.63	0.12	-8.31
	0.30	0.90	64	-0.58	0.14	-7.05
91	0.05	0.04	95	0.92	0.10	-9.75
	0.15	0.04	31	0.16	2.51	0.02
	0.30	0.08	38	-0.07	8.07	-0.20

* If all predicted and observed values were the same, then the statistics would yield: ME = 0.0; RMSE = 0.0; CRM = 0.0; CD = 1.0; and EF = 1.0.

Table X. Summary of 76 EDB simulations for D4201 with PRZM.

	Input Parameters†		1983								1985						
			Output		Statistics⁺					Output		Statistics				
Simulation Number*	Decay Rate@ (day-1)	Hydrodynamic Dispersion Coefficient# (cm2 day-1)	Depth to Peak Conc. (m)	Peak Conc. (µg/kg)	ME (µg/kg)	RMSE (%)	CRM	CD	EF	Depth to Peak Conc. (m)	Peak Conc. (µg/kg)	ME (µg/kg)	RMSE (%)	CRM	CD	EF
0	0.005000	13	7.4	8.96	10.68	86	0.62	1.01	-0.13	13.1	0.14	3.70	168	0.95	2.19	-0.41
1	0.010000	13	7.4	0.23	12.94	127	0.99	0.67	-1.47	13.1	0.00	3.80	173	1.00	2.00	-0.50
2	0.004700	13	7.4	11.16	10.10	80	0.52	0.96	0.02	13.1	0.22	3.64	165	0.93	2.30	-0.37
3	0.004400	110	10.8	10.22	8.70	63	0.16	1.93	0.40	17.6	6.05	6.03	365	-1.55	0.19	-5.64
4	0.004400	105	10.8	10.19	8.74	62	0.17	1.91	0.41	17.3	6.00	6.00	364	-1.54	0.19	-5.60
5	0.004400	100	10.7	10.15	8.79	62	0.17	1.89	0.42	17.1	5.97	5.97	362	-1.52	0.19	-5.56
6	0.004400	95	10.5	10.12	8.84	61	0.18	1.86	0.43	16.9	5.93	5.93	361	-1.51	0.19	-5.52
7	0.004400	90	10.5	10.09	8.88	61	0.19	1.82	0.43	16.7	5.91	5.89	360	-1.49	0.20	-5.47
8	0.004400	85	10.4	10.06	8.93	61	0.20	1.79	0.44	16.6	5.88	5.85	358	-1.47	0.20	-5.42
9	0.004400	80	10.3	10.03	8.99	60	0.21	1.74	0.45	16.4	5.86	5.80	357	-1.46	0.20	-5.36
10	0.004400	75	10.3	9.99	9.04	60	0.22	1.70	0.45	16.3	5.85	5.77	355	-1.44	0.20	-5.30
11	0.004400	70	10.2	9.96	9.09	60	0.24	1.65	0.46	16.2	5.84	5.79	353	-1.42	0.21	-5.24
12	0.004400	65	10.1	9.92	9.14	59	0.25	1.60	0.46	15.2	5.83	5.81	351	-1.39	0.21	-5.16
13	0.004400	60	10.0	9.87	9.19	59	0.26	1.55	0.47	15.2	5.87	5.84	349	-1.37	0.21	-5.08
14	0.004400	55	9.9	9.82	9.25	59	0.27	1.49	0.47	15.1	5.91	5.86	346	-1.34	0.22	-4.98
15	0.004400	50	9.8	9.76	9.30	59	0.28	1.43	0.47	15.1	5.95	5.88	343	-1.31	0.22	-4.87
16	0.004400	45	8.5	9.75	9.34	59	0.29	1.37	0.47	15.0	6.00	5.90	339	-1.27	0.23	-4.74
17	0.004400	40	8.5	9.92	9.39	59	0.30	1.30	0.47	14.9	6.05	5.91	335	-1.23	0.23	-4.60
18	0.004400	35	8.5	10.11	9.42	59	0.31	1.22	0.46	14.8	6.11	5.91	330	-1.18	0.24	-4.43
19	0.004400	30	7.5	10.64	9.45	60	0.32	1.14	0.45	14.7	6.17	5.90	324	-1.12	0.25	-4.23
20	0.004400	25	7.5	11.32	9.48	61	0.33	1.06	0.43	14.6	6.25	5.85	317	-1.05	0.26	-4.01
21	0.004400	20	7.4	12.20	9.49	63	0.33	0.96	0.39	14.5	6.33	5.82	309	-0.97	0.27	-3.77
22	0.004400	15	7.4	13.34	9.50	66	0.34	0.86	0.33	14.3	6.44	6.03	301	-0.87	0.29	-3.51
23	0.004400	13	7.8	24.54	16.01	104	-0.04	0.30	-0.65	13.9	12.77	12.47	557	-2.66	0.07	-14.48
24	0.004400	13	7.4	13.90	9.50	68	0.34	0.82	0.29	14.2	6.50	6.13	297	-0.83	0.29	-3.41

Table X continues.

Table X. Summary of 76 EDB simulations for D4201 with PRZM.

Simulation Number*	Input Parameters† Decay Rate@ (day-1)	Input Parameters† Hydrodynamic Dispersion Coefficient# (cm2 day-1)	1983 Output Depth to Peak Conc. (m)	1983 Output Peak Conc. (µg/kg)	1983 Statistics+ ME (µg/kg)	1983 Statistics+ RMSE (%)	1983 Statistics+ CRM	1983 Statistics+ CD	1983 Statistics+ EF	1985 Output Depth to Peak Conc. (m)	1985 Output Peak Conc. (µg/kg)	1985 Statistics ME (µg/kg)	1985 Statistics RMSE (%)	1985 Statistics CRM	1985 Statistics CD	1985 Statistics EF
25	0.004400	13	7.4	13.90	9.50	71	0.36	0.82	0.24	13.9	2.87	2.84	169	0.18	1.65	-0.42
26	0.004400	13	7.4	13.90	9.50	73	0.38	0.82	0.19	13.5	1.33	3.21	150	0.61	3.25	-0.13
27	0.004400	13	7.4	13.90	9.50	75	0.39	0.82	0.14	13.1	0.66	3.43	155	0.80	2.87	-0.21
28	0.004400	13	7.4	13.90	9.50	77	0.41	0.82	0.10	13.1	0.34	3.56	161	0.89	2.48	-0.30
29	0.004200	90	10.4	11.16	8.79	58	0.09	1.61	0.49	16.6	6.54	6.51	396	-1.79	0.15	-6.85
30	0.004200	85	10.4	11.14	8.85	57	0.10	1.58	0.50	16.5	6.52	6.47	395	-1.77	0.16	-6.80
31	0.004200	80	10.3	11.12	8.90	57	0.11	1.54	0.51	16.4	6.51	6.42	394	-1.76	0.16	-6.75
32	0.004200	75	10.2	11.09	8.96	56	0.13	1.50	0.51	16.2	6.50	6.44	392	-1.74	0.16	-6.69
33	0.004200	70	10.1	11.07	9.02	56	0.14	1.46	0.52	16.1	6.50	6.47	390	-1.72	0.16	-6.62
34	0.004100	90	10.4	11.74	8.73	57	0.04	1.47	0.50	16.6	6.88	6.85	417	-1.95	0.14	-7.68
35	0.004100	85	10.3	11.72	8.80	57	0.05	1.44	0.51	16.4	6.87	6.80	416	-1.94	0.14	-7.63
36	0.004100	80	10.2	11.71	8.86	56	0.06	1.41	0.52	16.3	6.86	6.77	414	-1.92	0.14	-7.58
37	0.004100	75	10.1	11.69	8.92	56	0.07	1.37	0.53	16.2	6.86	6.80	413	-1.90	0.14	-7.52
38	0.004100	70	10.0	11.67	8.99	55	0.08	1.33	0.53	15.2	6.87	6.84	411	-1.88	0.14	-7.45
39	0.004075	15	7.4	16.92	9.50	70	0.17	0.62	0.25	14.2	8.08	7.70	366	-1.37	0.17	-5.70
40	0.004075	13	7.4	17.63	9.57	80	0.25	0.60	0.02	13.1	0.55	3.41	154	0.82	2.78	-0.19
41	0.004000	95	10.4	12.36	8.61	58	-0.03	1.34	0.48	16.7	7.25	7.24	440	-2.14	0.12	-8.65
42	0.004000	85	10.2	12.34	8.74	57	-0.01	1.29	0.51	16.4	7.24	7.16	438	-2.11	0.12	-8.57
43	0.004000	75	10.1	12.33	8.88	56	0.01	1.23	0.53	16.2	7.24	7.19	435	-2.08	0.12	-8.46
44	0.004000	65	10.0	12.31	9.02	55	0.03	1.16	0.54	15.1	7.33	7.29	432	-2.04	0.13	-8.32
45	0.004000	55	8.5	12.38	9.16	55	0.06	1.07	0.54	15.0	7.47	7.38	428	-1.99	0.13	-8.14
46	0.004000	45	8.5	12.80	9.29	55	0.08	0.97	0.53	14.9	7.64	7.48	421	-1.93	0.13	-7.87
47	0.004000	35	7.6	13.50	9.40	57	0.10	0.85	0.50	14.7	7.85	7.54	412	-1.83	0.14	-7.50
48	0.004000	25	7.5	15.17	9.47	62	0.12	0.72	0.41	14.5	8.13	7.55	400	-1.70	0.15	-6.99
49	0.004000	15	7.4	17.88	9.84	73	0.13	0.56	0.19	14.2	8.52	8.14	385	-1.50	0.15	-6.39
50	0.003900	90	10.3	13.00	8.61	59	-0.08	1.17	0.48	16.5	7.63	7.57	462	-2.32	0.11	-9.68

Table X continues.

Table X. Summary of 76 EDB simulations for D4201 with PRZM.

	Input Parameters†		1983 Output		1983 Statistics+					1985 Output		1985 Statistics				
Simulation Number*	Decay Rate@ (day^{-1})	Hydrodynamic Dispersion Coefficient# (cm^2 day^{-1})	Depth to Peak Conc. (m)	Peak Conc. (µg/kg)	ME (µg/kg)	RMSE (%)	CRM	CD	EF	Depth to Peak Conc. (m)	Peak Conc. (µg/kg)	ME (µg/kg)	RMSE (%)	CRM	CD	EF
51	0.003900	85	10.2	13.00	8.69	58	-0.07	1.15	0.49	16.4	7.63	7.54	461	-2.30	0.11	-9.64
52	0.003900	80	10.1	13.00	8.76	57	-0.06	1.12	0.50	16.2	7.63	7.56	460	-2.29	0.11	-9.59
53	0.003900	75	10.0	13.00	8.83	57	-0.05	1.09	0.51	16.1	7.64	7.61	459	-2.27	0.11	-9.54
54	0.003900	70	10.0	13.00	8.91	56	-0.04	1.06	0.51	15.2	7.70	7.66	458	-2.25	0.11	-9.48
55	0.003900	65	9.9	13.00	8.98	56	-0.03	1.03	0.52	15.1	7.77	7.72	456	-2.23	0.11	-9.41
56	0.003900	60	8.5	13.04	9.06	56	-0.02	0.99	0.52	15.1	7.84	7.77	455	-2.21	0.11	-9.33
57	0.003900	55	8.5	13.24	9.13	56	-0.01	0.95	0.52	15.0	7.93	7.83	452	-2.18	0.11	-9.22
58	0.003900	50	8.5	13.45	9.21	56	0.01	0.91	0.51	14.9	8.02	7.88	450	-2.15	0.11	-9.10
59	0.003800	90	10.2	13.69	8.55	61	-0.15	1.02	0.44	16.4	8.04	7.97	488	-2.52	0.09	-10.88
60	0.003800	85	10.2	13.70	8.62	60	-0.14	1.00	0.45	16.3	8.05	7.94	487	-2.50	0.10	-10.85
61	0.003800	80	10.1	13.71	8.70	60	-0.13	0.98	0.46	16.2	8.06	7.99	486	-2.49	0.10	-10.81
62	0.003800	75	10.0	13.72	8.78	59	-0.12	0.95	0.47	15.2	8.08	8.05	485	-2.47	0.10	-10.76
63	0.003800	70	10.0	13.73	8.86	59	-0.11	0.93	0.47	15.1	8.15	8.11	484	-2.46	0.10	-10.71
64	0.003800	13	7.4	21.57	13.50	92	0.08	0.42	-0.30	13.1	0.81	3.22	146	0.74	3.15	-0.06
65	0.003750	13	7.4	22.37	14.30	95	0.04	0.40	-0.39	13.1	0.88	3.18	144	0.72	3.22	-0.04
66	0.003500	95	10.2	15.99	8.22	75	-0.39	0.66	0.15	16.4	9.42	9.34	576	-3.22	0.07	-15.55
67	0.003500	85	10.0	16.06	8.41	74	-0.37	0.63	0.16	16.2	9.46	9.40	576	-3.20	0.07	-15.56
68	0.003500	75	8.5	16.32	8.61	74	-0.34	0.60	0.17	15.1	9.61	9.56	576	-3.18	0.07	-15.55
69	0.003500	65	8.5	16.77	8.81	74	-0.32	0.57	0.16	15.0	9.83	9.73	575	-3.15	0.07	-15.50
70	0.003500	55	8.5	17.32	9.01	75	-0.30	0.53	0.13	14.8	10.10	9.92	573	-3.11	0.07	-15.40
71	0.003500	45	8.5	17.99	9.75	78	-0.28	0.48	0.06	14.7	10.43	10.11	569	-3.06	0.07	-15.20
72	0.003500	35	7.5	19.43	11.43	84	-0.26	0.42	-0.07	14.6	10.85	10.27	563	-2.97	0.07	-14.86
73	0.003500	25	7.5	21.87	13.87	94	-0.24	0.35	-0.34	14.4	11.39	10.86	555	-2.83	0.07	-14.38
74	0.003500	15	7.4	25.79	17.73	111	-0.24	0.27	-0.90	14.0	12.19	11.86	545	-2.61	0.07	-13.86
75	0.002500	13	7.4	55.95	47.78	307	-1.39	0.05	-13.40	13.1	5.46	4.50	185	-0.77	0.42	-0.72

+ If all predicted and observed values were the same, then the statistics would yield: ME=0.0; RMSE=0.0; CRM=0.0; CD=1.0; and EF=1.0.
* The initial case (0) is Case H from Loague et al. (1989c); Cases 1-75 constitute a trial-and-error calibration.
† All other input parameters are the same as given by Loague et al. (1989b) and Loague et al. (1989c); the fifth realization of rainfall scheme III (Loague et al., 1989c) is used for all 76 simulations.
@ 0.0 (8.5 - 20 m) for #'s 3-24, 29-39, 41-63, 66-74; 0.0022 (5.5 - 8.5 m) for #23; 0.0033 (8.5 - 20 m) for #25; 0.0022 (8.5 - 20 m) for #26; 0.0011 (8.5 -20 m) for #27.
# Not adjusted for numerical dispersion.

Table XI. Summary of ranked PRZM simulations. The evaluations are based upon concentration profile characteristics.

	1983			1985			
Simulation Number	Depth to Peak Conc.	Peak Conc.	Average	Depth to Peak Conc.	Peak Conc.	Average	Total Average
0	20	36	14	1	45	20	10
1	20	70	61	1	47	22	42
2	20	6	5	1	44	15	5
3	75	8	47	76	16	52	57
4	75	9	48	75	15	50	56
5	74	11	52	74	13	43	54
6	72	12	48	73	11	40	48
7	72	14	53	71	9	36	49
8	67	15	46	68	8	29	39
9	62	16	40	59	6	26	32
10	62	17	43	56	5	25	35
11	54	18	30	49	4	24	26
12	48	20	26	42	3	15	18
13	40	21	16	42	7	23	14
14	38	24	17	35	10	15	8
15	37	28	22	35	12	21	20
16	1	29	6	31	14	15	6
17	1	19	3	28	17	15	4
18	1	13	1	26	18	14	3
19	15	1	2	23	19	12	1
20	15	10	4	21	20	10	1
21	20	30	12	19	22	10	7
22	20	46	23	17	23	9	8
23	13	74	56	11	76	43	57
24	20	54	33	14	25	7	17
25	20	54	33	11	1	2	11
26	20	54	33	10	21	3	12
27	20	54	33	1	37	6	14
28	20	54	33	1	42	13	22
29	67	6	32	68	29	59	51
30	67	5	30	66	28	54	43
31	62	4	23	59	27	41	31
32	54	3	15	49	26	28	20
33	48	2	12	47	24	27	14
34	67	27	67	68	36	66	69
35	62	26	60	59	35	54	63
36	54	25	43	56	33	48	51

Table XI continues.

Table XI. Summary of ranked PRZM simulations. The evaluations are based upon concentration profile characteristics.

	1983			1985			
Simulation Number	Depth to Peak Conc.	Peak Conc.	Average	Depth to Peak Conc.	Peak Conc.	Average	Total Average
37	48	23	28	49	32	38	32
38	40	22	17	42	34	29	22
39	20	64	48	14	63	31	40
40	20	66	53	1	38	7	29
41	67	34	72	71	41	71	71
42	54	33	56	59	39	61	65
43	48	32	45	49	39	45	50
44	40	31	28	35	43	33	30
45	1	35	7	31	46	31	13
46	1	37	8	28	51	34	19
47	14	48	17	23	56	34	24
48	15	59	33	19	64	39	38
49	20	67	56	14	66	36	53
50	62	38	71	66	49	72	71
51	54	38	63	59	48	69	68
52	48	38	53	49	50	63	64
53	40	38	40	47	51	61	59
54	40	38	40	42	53	57	55
55	38	38	39	35	54	48	47
56	1	44	9	35	55	50	28
57	1	45	10	31	57	45	27
58	1	47	11	28	58	41	25
59	54	49	74	59	59	75	75
60	54	50	75	56	60	73	74
61	48	51	70	49	61	70	70
62	40	52	63	42	62	66	66
63	40	53	65	35	65	64	66
64	20	71	62	1	31	5	34
65	20	73	65	1	30	3	35
66	54	60	76	59	67	76	76
67	40	61	72	49	68	74	73
68	1	62	20	35	69	66	45
69	1	63	21	31	70	65	45
70	1	65	23	26	71	59	41
71	1	68	27	23	72	57	43
72	15	69	48	21	73	54	60
73	15	72	56	18	74	52	61
74	20	75	68	13	75	45	62
75	20	76	69	1	2	1	37

Table XII. Summary of ranked PRZM simulations. The evaluations are based upon statistical criteria.

Simulation Number	1983						1985						Total Average
	ME	RMSE	CRM	CD	EF	Average	ME	RMSE	CRM	CD	EF	Average	
0	68	69	74	3	69	64	9	8	12	69	8	24	62
1	70	75	75	35	75	74	10	10	14	68	10	26	66
2	67	67	73	9	67	64	8	7	11	70	7	22	57
3	9	50	37	75	50	41	30	32	34	23	32	32	42
4	13	48	38	74	48	41	29	31	33	22	31	31	40
5	17	47	40	73	47	45	28	30	32	21	30	30	45
6	21	45	41	72	45	45	27	29	31	20	29	29	44
7	26	44	42	71	44	48	22	28	29	19	28	28	48
8	30	42	43	70	42	48	18	27	28	18	27	27	45
9	33	41	44	69	41	50	14	26	27	17	26	25	45
10	38	38	45	67	38	47	12	25	26	16	25	23	37
11	40	36	46	65	36	44	13	24	25	15	24	21	30
12	42	34	49	61	34	40	15	23	24	14	23	19	19
13	44	32	51	57	32	39	17	22	23	13	22	18	16
14	46	30	53	53	30	36	20	21	21	12	21	16	7
15	48	27	54	46	27	33	21	20	20	11	20	14	5
16	49	28	56	39	28	32	24	19	19	10	19	13	3
17	50	31	57	34	31	34	25	18	18	9	18	12	4
18	52	35	59	30	35	35	26	17	17	8	17	9	1
19	53	40	60	20	40	37	23	16	16	7	16	7	1
20	55	46	62	12	46	41	19	15	15	6	15	6	5
21	56	51	63	8	51	51	16	14	13	5	14	3	9
22	58	52	64	19	52	55	30	13	9	4	13	5	22
23	74	73	11	66	73	70	76	69	68	64	69	70	72
24	59	53	65	27	53	58	32	12	8	3	12	4	25

Table XII continues.

Table XII. Summary of ranked PRZM simulations. The evaluations are based upon statistical criteria.

Simulation Number	1983						1985						Total Average
	ME	RMSE	CRM	CD	EF	Average	ME	RMSE	CRM	CD	EF	Average	
25	59	55	67	25	55	60	1	9	1	2	9	1	23
26	59	57	69	26	57	61	3	3	2	76	3	11	40
27	59	62	71	28	62	63	6	5	6	73	5	16	54
28	59	64	72	29	64	66	7	6	10	71	6	20	57
29	16	24	24	63	24	31	37	39	40	29	39	38	36
30	22	21	26	58	21	30	35	38	39	28	38	37	33
31	27	18	28	56	18	29	33	37	38	27	37	36	30
32	31	14	31	54	14	28	34	36	37	26	36	35	26
33	37	8	34	50	8	26	36	35	36	25	35	33	19
34	11	19	10	51	19	19	42	46	47	37	46	46	30
35	18	15	13	49	15	19	40	45	46	35	45	44	26
36	23	11	16	45	11	18	38	44	44	34	44	43	23
37	29	5	18	41	5	14	39	43	43	33	43	42	13
38	34	3	22	36	3	14	41	41	42	32	41	40	9
39	57	54	39	42	54	57	56	33	22	24	33	34	63
40	64	66	50	44	66	67	5	4	7	72	4	14	55
41	5	25	7	37	25	16	45	52	52	43	52	52	34
42	12	16	3	33	16	8	43	51	51	42	51	51	19
43	25	6	4	31	6	4	44	50	50	41	50	50	9
44	36	2	8	23	2	3	46	49	49	40	49	49	7
45	43	1	15	14	1	6	47	48	48	39	48	48	9
46	47	4	20	7	4	9	48	47	45	38	47	47	13
47	51	22	25	22	22	27	50	42	41	36	42	44	38
48	54	49	29	32	49	37	51	40	35	31	40	40	50
49	66	56	32	48	56	59	66	34	30	30	34	39	65

Table XII continues.

Table XII. Summary of ranked PRZM simulations. The evaluations are based upon statistical criteria.

Simulation Number	1983						1985						Total Average
	ME	RMSE	CRM	CD	EF	Average	ME	RMSE	CRM	CD	EF	Average	
50	6	26	23	24	26	17	53	61	61	52	61	61	51
51	8	23	19	21	23	13	49	60	60	51	60	60	42
52	14	20	17	18	20	12	52	59	59	50	59	59	38
53	20	17	14	16	17	10	54	58	58	49	58	58	34
54	28	13	9	13	13	7	55	57	57	48	57	57	28
55	32	10	6	6	10	2	57	56	56	47	56	56	17
56	39	7	5	2	7	1	58	55	55	46	55	55	13
57	41	9	2	11	9	4	59	54	54	45	54	54	17
58	45	12	1	17	12	11	60	53	53	44	53	53	28
59	3	43	36	4	43	25	62	66	66	57	66	66	63
60	7	39	35	1	39	21	61	65	65	56	65	65	57
61	10	37	33	5	37	23	63	64	64	55	64	64	61
62	15	33	30	10	33	21	64	63	63	54	63	63	56
63	24	29	27	15	29	24	65	62	62	53	62	62	57
64	71	70	21	59	70	68	4	2	4	74	2	10	51
65	73	72	12	62	72	68	2	1	3	75	1	8	48
66	1	61	70	38	61	54	67	74	76	64	74	74	68
67	2	60	68	40	60	53	68	76	75	67	76	76	70
68	4	58	66	43	58	51	69	75	74	66	75	75	67
69	19	59	61	47	59	55	70	73	73	63	73	73	68
70	35	63	58	52	63	62	71	72	72	62	72	72	71
71	65	65	55	55	65	71	72	71	71	61	71	70	73
72	69	68	52	60	68	72	73	70	70	60	70	69	73
73	72	71	48	64	71	73	74	68	69	59	68	68	73
74	75	74	47	68	74	75	75	67	67	58	67	67	76
75	76	76	76	76	76	76	11	11	5	1	11	2	51

References

Carsel, R.F., C.N. Smith, L.A. Mulkey, J.D. Dean, and P. Jowise, 1984: *User's manual for the pesticide root zone model (PRZM), release 1.* US/EPA-600/3-84-109. U.S. Environmental Protection Agency, Washington, D.C.

Gilmour, P. 197: *General validation procedure for computer simulation models.* Australian Computer J., **5**, 127-130.

Green, I.R.A. and D. Stephenson, 1986: *Criteria for comparison of single event models.* Hydrological Sciences J., **31**, 395-411.

James, L.D. and S.J. Burges, 1982: *Selection, calibration, and testing of hydrologic models.* pp. 437-472. In: C.T. Haan, H.P. Johnson, and D.L. Brakensiek (eds.), Hydrologic Modeling of Small Watersheds. American Society of Agriculture Engineers, St. Joseph, Michigan.

Khan, M.A. and T. Liang, 1989: *Mapping pesticide contamination potential.* Environmental Management, **13**, 233-242.

Kleveno, J.J., 1990: *Evaluation of a pesticide mobility index used to assess the potential for groundwater contamination.* M.S. thesis. University of Hawaii, Manoa.

Loague, K. and R.E. Green, 1990a. *Statistical and graphical methods for evaluating solute transport models: Overview and application.* J. Contaminant Hydrology (in press).

Loague, K. and R.E. Green. 1990b: *Comments on "Mapping pesticide contamination potential"* by M. Akram Khan and Tung Liang. Environmental Management (in press).

Loague, K.M., R.S. Yost, R.E. Green, and T.C. Liang, 1989a: *Uncertainty in a pesticide leaching assessment for Hawaii.* J. Contaminant Hydrology, **4**, 139-161.

Loague, K.M., R.E. Green, C.C.K. Liu, and T.C. Liang, 1989b: *Simulation of organic chemical movement in Hawaii soils with PRZM: 1. Preliminary results for EDB.* Pacific Science, **43**, 67-95.

Loague, K., T.W. Giambelluca, R.E. Green, C.C.K. Liu, T.C. Liang, and D.S. Oki., 1989c: *Simulation of organic chemical movement in Hawaii soils with PRZM: 2. Predicting deep penetration of DBCP, EDB, and TCP.* Pacific Science, **43**, 362-383.

Loague, K., R.E. Green, T.W. Giambelluca, T.C. Liang, and R.S. Yost, 1990: *Impact of uncertainty in soil, climatic, and chemical information in a pesticide leaching assessment.* J. Contaminant Hydrology (in press).

Martinec, J. and A. Rango, 1989: *Merits of statistical criteria for the performance of hydrological models.* Water Resour. Bull., **25**, 421-432.

Peterson, F.L., K.R. Green, R.E. Green, and J.N. Ogata, 1985: *Drilling program and pesticide analysis of core samples from pineapple fields in central Oahu.* Water Resources Research Center, University of Hawaii at Manoa, Spec. Rept., 7.5 (unpublished).

Rao, P.S.C., A.G. Hornsby, and R.E. Jessup, 1985: *Indices for ranking the potential for pesticide contamination of groundwater.* Soil Crop Sci. Soc. Florida Proc., **44**, 1-8.

Smith, C.M., G.W. Bailey, R.A. Leonard, and G.W. Langdale, 1978: *Transport of agricultural chemicals from small upland piedmont watersheds.* EPA-6001 3-78-056. U.S. Environmental Protection Agency, Athens, Georgia.

Tichelaar, B.W. and L.J. Ruff, 1989: *How good are our best models? Jackknifing, bootstrapping, and earthquake depth.* EOS 70, **593**, 605-606.

Willmott, C.J., S.G. Acklenson, R.E. Davis, J.J. Feddema, K.M. Klink, D.R. Legates, J. O'Donnell, and C.M. Rowe, 1985: *Statistics for the evaluation and comparison of models.* J. Geophysical Research, **90** (C5), 8995-9005.

Wong, L., 1983: *Preliminary report on soil sampling EDB on Oahu.* Pesticide Branch, Div. of Plant Industry, Dept. of Agriculture, State of Hawaii (unpublished).

Wong, L., 1987: *Analysis of ethylene dibromide distribution in the soil profile following shank injection for nematode control in pineapple culture.* pp. 28-40. In: P.S.C. Rao and R.E. Green (eds.), Toxic Organic Chemicals in Hawaii's Water Resources Hawaii Inst. Trop. Agric. Hum. Resources Res. Exten. Ser. 086. University of Hawaii, Honolulu.

Keith Loague, Department of Soil Science, 108 Hilgard Hall, University of California, Berkeley, CA 94720, USA.
R.E. Green, Agronomy and Soil Science, University of Hawaii, Honolulu, HI, USA

Field-Scale Water and
Solute Flux in Soils
Monte Verità

RELATING THE PARAMETERS OF A LEACHING MODEL TO THE PERCENTAGES OF CLAY AND OTHER SOIL COMPONENTS

T.M. Addiscott and N.J. Bailey

The model divides the soil into layers and considers two categories of water, mobile and immobile, in each layer. It has two main parameters, one a measure of the soil's capacity to hold water and solutes and the other a measure of the ease with which water can pass through the soil and carry solutes with it. Both parameters can be derived from the percentages of clay and other soil components. The model is shown to be fairly robust with respect to variability in its parameters, and the relationships by which the parameters are estimated from the percentages of clay etc. are also not seriously affected by soil variability. Using the model to map leaching in a site with two soil types gave a result that seemed to correspond with the soil survey map.

1. Introduction

Those who model the movement of solute and water through the soil face an awkward problem. The more rigorous a model is in mechanistic and mathematical terms the less likely it is that all its parameters can be known with certainty. There are several possible responses to this problem. One is to use a mechanistic model and to recognise the uncertainty by treating its parameters as probability distributions (e.g. Dagan and Bresler, 1979; Amoozegar-Fard *et al.*, 1982). Another is to assume that the uncertainty is the dominant feature of the system and use a very simple transfer function model in a stochastic framework (Jury, 1982; Jury *et al.*, 1982). A third alternative is to use a less mechanistic model whose parameters can be known with greater certainty; this is the approach that we adopt. Models of this kind, classified as 'functional' by Addiscott and Wagenet (1985a), tend to be used for a specific purpose, usually the management of the solute. This makes it important that the model's parameters should be able to be derived from independent information — as is true of most models.

Another distinction in the classification cited above was between models that depend primarily on *rate* parameters, such as the hydraulic conductivity, and those that use *capacity* parameters that usually derive from the volumetric moisture content. Rate parameters are much more subject to spatial variability than capacity parameters. For example, Jones and Wagenet (1984) found the *logarithm* of the saturated hydraulic conductivity to have a coefficient of variation (CV) of 71% while the CV of the untransformed volumetric moisture content was 14%. A model that depends on capacity parameters alone should be little affected by their variability but will probably not discriminate satisfactorily between soils of differing textures. A model that depends on rate parameters alone should, on the other hand, discriminate between soils but suffer much more from the variability of its parameters. Perhaps, then, we need a model with both rate and capacity parameters, the former to discriminate between soils of differing textures and the latter to stabilize the model against the variability of the former. Hence the model to be described, acronym SLIM (Solute Leaching Intermediate Model).

The uncertainty mentioned above results from the variability that is found when soil properties are measured at a number of points within a site. This is but one aspect of the overall heterogeneity of the soil. We have also to consider small-scale structural heterogeneity and, at the other end of the scale, the existence of soil types with clearly differing general properties. 'Heterogeneous' means 'diverse in character' or made up of several component parts (or elements), and it is the soil's component parts, clay, sand, silt, organic matter, water and air that are the key to all aspects of its heterogeneity. They are responsible for the aggregation that leads to the small-scale structural heterogeneity. Variations in their relative proportions contribute importantly to variations in soil properties. And the diverse soil types are characterized largely by their composition with respect to these soil components. These components have a strong influence on the soil's ability to hold or to transmit water, and the holding and transmission of water are central to any consideration of solute leaching. This paper examines the consequences of soil variability for the functioning of the SLIM model and the derivation of its parameters.

2. The Model and its Parameters

The small-scale structural heterogeneity mentioned above implies that even in a very small volume of soil there will be widely differing rates of water movement. This concept, first discussed by Lawes *et al.* (1882), is usually simplified by assuming that there are mobile and immobile categories of water in the soil, which for many soils broadly means water outside or inside soil aggregates or larger structural units. The highly irregular breakthrough curves found for chloride in the Rothamsted Drain Gauges (lysimeters) provide strong circumstantial evidence for this concept (Addiscott *et al.*, 1978). Immobile water is not, however, a feature of aggregated soils alone, since De Smedt *et al.* (1986) found evidence of it in a column of pure sand. Solutes are safe from leaching as long as they remain in immobile water, and the 'mobile-immobile' concept of soil water has been incorporated in a number of leaching models (e.g. Passioura, 1971; Van Genuchten and Wierenga, 1976; Addiscott, 1977).

The model that is the focus of this paper is a layer model that divides the soil water into mobile and immobile categories W_m and W_r, each of which may contain solute, S_m and S_r respectively, where m denotes the mobile category and r the immobile (or retained) category. As in an earlier model (Addiscott, 1977), W_r is fixed in the sense that it is undiminished by drainage, although it may be depleted by evaporation. However, its definition in this model is in terms of the volumetric moisture content of the soil at 0.33 bars rather than 2 bars as before; this change and other details of the new model are discussed in a paper shortly to be submitted for publication (Addiscott and Whitmore, 1991).

$$d\left(\theta_{0.33}-\frac{1}{2}\theta_{15}\right)\geq W_r\geq 0 \qquad (1)$$

As before the θ_{15} that is subtracted from $\theta_{0.33}$ represents the most strongly-bound water which is assumed to be inaccessible to anionic solutes because of anion exclusion. Each layer has a thickness *d*.

Water and solute entering a given layer from the layer above or from rainfall are added to the current W_m and S_m respectively. A proportion, θ of the new W_m and S_m then moves to the next layer to be added to the W_m and S_m there, and so on. As in the previous model, solute moves between the mobile and immobile water,but the present model is structured such that half the vertical movement of water and solute occurs before this lateral solute movement and half after. This is to take account of the fact that lateral solute movement must occur in reality during, as well as after, the vertical flow in the mobile water. The lateral movement is by diffusion and this can be treated in three ways. (a) The solute concentrations can simple by equalized between W_m and W_r; this implies that diffusion is completed, which cannot happen in finite time. (b) Diffusion can be simulated explicitly if suitable assumptions are made about the geometry of the soil structure (Addiscott, 1982). (c) The limits imposed by diffusion can be described by partially equalising concentrations between W_m and W_r using a 'hold-back'factor β to limit the degree of equalization. The latter procedure was used here. It should be noted that α is not independent of the layer thickness, so that latter is standardised at 50 mm.

The model has an evaporation routine in which open-water evaporation is moderated according to the dryness of the soil. Evaporative demand is initially met from the top layer and the resulting deficit is then shared with the second and subsequent layers as in the CALF model of Nicholls *et al.* (1982). The water and solute movement associated with the deficit sharing are assumed to occur through the finer pores in which the immobile water is found.

The two main parameters of the model are W_r and α. These have contrasting roles; W_r determines the soil's capacity for holding back solute against leaching, which α is a simple permeability parameter that measures the ease with which water flow can occur to remove solute by leaching. The nature of W_r is clear from the way in which it is derived from the soil moisture characteristic, but α needs further elaboration. It can be estimated by generating simulated profiles of the volumetric moisture content, θ_v, from a series of α-values and finding the simulated profile that best matches the measured profile. This is done for soils under natural rainfall and evaporation over a period of several months in the field. If α is properly to be regarded as a permeability parameter it should be related to some other measure of soil permeability, and this was tested as follows. The soil in a 50 m x 50 m plot was sampled with a 50 mm auger at 36 randomly distributed points to 1.0 m in 0.2 m increments in autumn and spring, and α established at each point from the changes in the profile of θ_v. The holes left were filled with water to the level of the soil surface and the fall in level, h, in the first hour was measured to give a coarse estimate of the permeability. Three approaches were used to assess how well α was related to h, taking the autumn values for h.

1) *Correlation.* This gave a highly significant positive correlation coefficient. $r=0.49$ ($p<0.01$).
2) *Spearman's ranking test.* The ranking coefficient, 0.53 ($t=3.62$; $p<0.001$) was significant at a higher level than the correlation coefficient.
3) *Mapping α and h.* Contour diagrams produced using UNIMAP which is part of the UNIRAS graphics package showed generally similar patterns (Figure 1) with several spatial features, notably A, B and C, common to both varieties.

These results strongly suggest that α and h were measures of the same category of soil property.

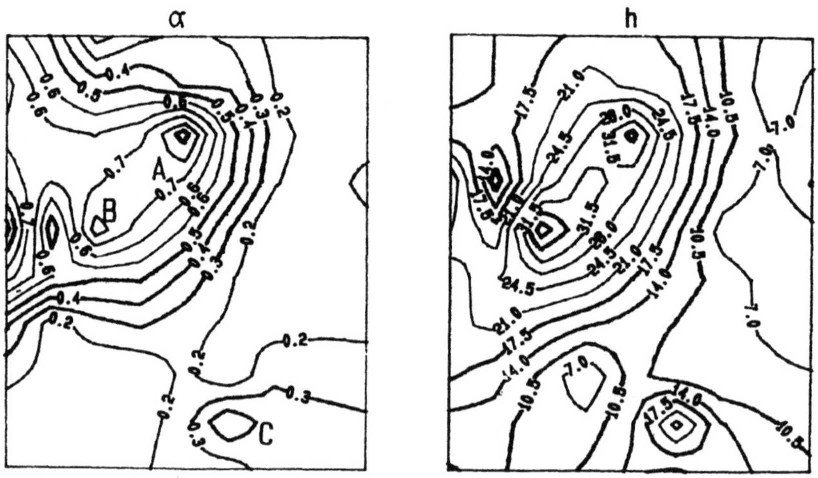

Figure 1: Contour diagrams of α and h.

3. Obtaining values for the parameters

A model whose parameters have to be obtained by fitting the model to the data to be simulated is less useful than one whose parameters can be obtained independently. The parameter W_r of the SLIM model can be obtained from the soil moisture characteristic (Equation (1)), but, where there is a clear distinction between the topsoil and the subsoil separate values of W_r for each are desirable. If separate soil moisture characteristics are not available the parameter's dependence on the soil's composition can be utilised. Hall *et al.* (1977) used data from a large number of British soils to compute the regressions of the volumetric moisture content, θ at various tensions on the % clay (C), % silt (Z), % organic carbon (X) and bulk density (D_b). We are concerned with θ-values defining the upper and lower limits of W_r, $\theta_{0.33}$ and θ 15 respectively. Hall *et al.* did not compute a regression for $\theta_{0.33}$ but they did for $\theta_{0.4}$ (i.e. at θ at 0.4 bars) and since the difference between $\theta_{0.33}$ and $\theta_{0.4}$ is likely to be negligible the latter is shown. Separate regressions were given for topsoils and subsoils and the θ-values that the yield are percentages rather than decimal fractions.

Topsoils

$$\theta_{0.4} = 26.66 + 0.36C + 0.12Z + 1.00X - 7.64D_b \quad (2)$$

$$\theta_{15} = 2.94 + 0.83C - 0.0054C^2 \quad (3)$$

Subsoils

$$\theta_{0.4} = 20.81 + 0.45C + 0.13Z + 5.96D_b \quad (4)$$

$$\theta_{15} = 1.48 + 0.84C + 0.0054C^2 \quad (5)$$

These regressions accounted for 70, 73, 77 and 83 percent of the variation respectively and they provide a second means by which W_r can be estimated independently of the data to be simulated.

The permeability parameter α was initially obtained by fitting the model to moisture content profiles. Using the resulting values of α to simulate solute movement is arguably valid because the model was not fitted to the solute data, but a more independent means of estimating α would be preferable. Here again we turn to the clay, sand and silt in the soil which are known to influence the soil's permeability to water (Derr *et al.*, 1969, Winneburger, 1974, Jaynes and Tyler, 1984). Estimates of α were made as described above from soil moisture profiles measured by the senior author and other members of Rothamsted in thirteen soils containing from 5 to 56 percent clay, some of which were sampled twice. (This % clay is the average figure for the profile because α relates to the whole profile.) The results showed α to be correlated negatively with the % clay ($r=-0.91$; $p<0.001$) and positively with the % sand ($r=0.72$; $p<0.001$) but uncorrelated with the % silt ($r=0.001$). The relationship with % clay showed two deviant soils. One was a peat containing 22% organic matter by weight and more by volume; since the % clay was determined after ignition it was meaningless for estimating α. For this soil α was 1.0. The other soil was not deviant for any obvious reason, but omitting it halved the percentage of variation not accounted for in the best relationship found for α, that with the square of the % clay (C). With the peat and the other deviant omitted, the regression of α on C^2,

$$\alpha = 1.0271 - 0.00302C^2 \quad (6)$$

accounted for 94% of the variation. Since α cannot by definition exceed 1.0 or be less than 0.0, Equation(6) must be re-expressed.

$$9.5 < C > 58.3 \quad \alpha = 1.0271 - 0.000302C^2 \quad (6a)$$

$$9.5 < C > 58.3 \ \alpha = 1.0271 - 0.000302C^2 \quad (6b)$$

$$C \geq 58.3 \ \ \alpha = 0.0 \quad (6c)$$

It should be noted that this relationship was derived from α-values measured during winter when the soil was fully moist. If the soils with the larger clay percentages had been dry enough for appreciable cracking to have persisted, the relationship would presumable have been somewhat different. This is part of the overall problem that the relation does not take account of the effects on conductivity of structural features such as voids in clay soils (Anderson and Bouma, 1973).

4. The Impact on the Model of Variability within a Soil Type

4.1 Variability in the Parameters

We assessed the impact of variability in the model's parameters on its performance by finding out how much simulation of some field leaching data was improved by taking account of the variability (Addiscott and Bland, 1988). The experiment, in which nitrate was applied to the surface of twelve 12 m x 2 m field plots on November 1 and allowed to leach under natural rain and evaporation, was similar to one reported earlier (Addiscott *et al.*, 1986) except that the soil was sampled to a greater depth, 1.0 m, in 0.2 m increments. All 12 plots were sampled for nitrate and soil moisture, each at six points, with a 20 mm semi-cylindrical auger; samples for each depth increment were bulked within plots. Six further plots, treated identically except that no nitrate was given, provided a control that enabled the percentage recovery of nitrate to be calculated by difference. Separate values of W_r for the topsoils (t) and the subsoil (s) were considered; α is for the whole profile. The means and standard deviations (SD) of the parameters were as follows.

	W_r(t) (mm)	W_r(s) (mm)	α
Mean	9.1	11.6	0.61
SD	0.46	0.81	0.21
CV (%)	5	7	35

Thus α was very much the most variable. Its distribution was also the most skewed, but it was not clearly normally or log-normally distributed.

The percentages of the applied nitrate in each layer of the profile at samplings on December 3 and February 29 were simulated fairly satisfactorily by the model using the means of the parameters without reference to the variability about the means (Figure 2). Taking account of the variability meant using the probability distributions of the parameters in the model. This involved using a procedure described as the 'Sectioning Method' (Addiscott and Wagenet, 1985b) in which the probability distribution of each parameter is divided into sections, each corresponding to the same number of observations. The section medians are then combined through the model in all combinations of section and parameter to generate a population of values for each output variate, without presupposing any particular form of distribution for it, and the moments of an appropriate distribution can then be computed. Simulations were made with all the parameter variances omitted (as in Figure 2), all of them included, or with the variances of individual parameters omitted. Two criteria were used to assess the agreement between measured and simulated profiles of applied nitrate: the overall percentage recovery of the nitrate, and a procedure (Whitmore, in preparation) that tested the simulated recoveries of nitrate in the five depth increments against the means and variances of the measured recoveries and attributed a mean square to lack of fit (LOF).

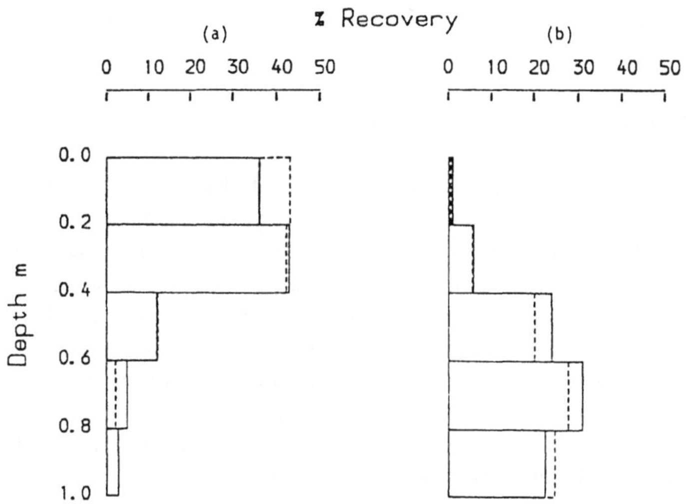

Figure 2: Recovery of nitrate at first and second samplings, by layers.
——— *measured,* - - - - - *simulated*

The total measured recovery of applied nitrate at the first sampling was 98.8% and the simulated recoveries ranged from 99.3% to 99.9% (Table I) giving little scope for comparison. Nearly all the applied nitrate was in the top two sampling layers and all the simulations showed too much nitrate in the top layer. The smallest LOF mean square (best fit) was given by the simulation that used the means of $W_r(t)$, $W_r(s)$ and α without their variances. Thus at this stage there was no evidence that ignoring within-site variability lessened the quality of the simulation.

By the second sampling, there had been much more leaching and the nitrate peak had reached the fourth sampling layer (0.6-0.8 m). Any errors arising from the omission of the variances would have accumulated to a much greater extent, and this time the inclusion of the variances of $W_r(t)$, $W_r(s)$ and α appreciably improved the simulation of total recovery and decreased the LOF

Table I: Effects of parameter variances on simulations of nitrate recovery

Simulation	Sampling 1 Total rec (%)	Sampling 1 LOF Mean Sq.	Sampling 2 Total rec (%)	Sampling 2 LOF Mean Sq.
(Measured)	98.8	—	84.2	—
All σ^2 omitted	99.3	95	78.4	47.2
All σ^2 included	99.9	173	83.6	11.4
σ^2_α omitted	99.9	136	83.4	25.9
σ Wr(t) omitted	99.9	176	83.7	11.6
σ^2 Wr(s) omitted	99.9	173	83.8	11.1

mean square (improved the fit) by a factor of four (Table I). Omitting the variances of $W_r(t)$, $W_r(s)$ had little effect on the LOF mean square but omitting that of α increased it, i.e. the most important variance to include seemed to be that of α, presumably because α had the largest coefficient of variation. In the context of practical solute management simulating a recovery of 78% rather than 84% of the nitrate would usually be acceptable. The difference is certainly no larger, for example, than the likely difference between a farmer's intended and actual applications of nitrogen fertilizer. We can conclude, with respect to this experiment at least, that the impact of the variation in parameters on the SLIM model was readily explicable and not particularly important. The model also gave, with single parameter values, satisfactory simulations of the chloride leaching data of Burns (1974) and Cameron and Wild (1982).

4.2 Variability in the Estimators of the Parameters

The clay, sand, silt and organic matter in a soil all vary in their relative proportions and we need to check that this does not effect the reliability of Equations (2), (3), (4), (5) and (6) for estimating $W_r(t)$, $W_r(s)$ and α. Equations (2) and (4) are linear with respect to all their variables and offer not problem. Equations (3), (5) and (6), however, include C^2, so variability in C, the percentage of clay, will introduce errors when these equations are used. These can readily be amended using a correction term given by Rao *et al.* (1977), and the corrections are not large. For example, calculations based on the measurements made by Avery and Bullock (1969) in Broadbalk field at Rothamsted suggest that the coefficients of variation of the percentages of clay in the topsoil, subsoil and whole profile are about 15, 37 and 31 percent respectively. These translate into corrections of:

0.05 mm in $W_r(t)$, i.e. about 0.5%
0.6 mm in $W_r(s)$, i.e. about 5%
0.04 in α, i.e. about 6%

The percentage corrections to $W_r(t)$ and α are much smaller than the coefficients of variability quoted earlier; that to $W_r(s)$ is comparable with the cv. Thus variability in the % clay should not be a serious problem; the correction is fairly small and easily made.

5. Mapping Leaching where there is more than One Soil Type

Different parts of a field may need differing amounts of fertilizer, due to variations in soil type or the history of the soil. If these variations are mapped and the map programmed into a computer on a special tractor with a fertilizer spreader, the quantity of fertilizer supplied can be varied appropriately, as described by Mulla and Hammond (1988) and Veseth (1989). Maps showing variations in leaching losses could be very useful. We explained the possibility of such a map for a 50 m x 50 m site at Rothamsted for which the Soil Survery map shows two soil types, one of which is described as the "eroded" form of the other because it has lost loess from the topsoil and therefore has the grater percentage of clay in the topsoil. This was the site on which the 36 values of a discussed in Section 2 were obtained. These α-values proved to be bi-modally distributed when various distributions were tested using the Maximum Likelihood Programme of (Ross 1980), and this suggests two soil types. The model was run for each of the 36 points with the appropriate values of $W_r(t)$, $W_r(s)$ and α and with natural rain and evaporation to simulate the leaching of a non-adsorbed solute applied to the soil surface. The outputs from the model were the percentages of the solute recovered in the topsoil (0-0.25 m) and the subsoil (0.25-1.0 m), S_T and S_S respectively, and the percentage lost beneath 1.0 m, S_D. These outputs were interpolated on to an 8 x 9 grid with the UNIMAP package, which uses bilinear interpolation, quadratic interpolation and smothing, and mapped from the grid cells. UNIMAP is part of the UNIRAS graphics package. The pattern of S_T values bore a fairly clear relationship to the soil map (Figure 3), while the patterns for S_S and S_D (not shown) showed some relationship but a much less clear one. This is not surprising because the differences between the two soil types lie mainly in the topsoil.

The above results were obtained by running the model for each point at which its parameters were estimated and interpolating and mapping the output. We wondered if interpolating the model's parameters on to the grid and running the model for each grid cell would give the same results when the outputs were mapped from the grid. The countour diagrams for S_T obtained from the two procedures showed considerable similarities (Figure 3), but the corresponding diagrams for S_S (not shown) were not at all similar. There was a greater degree of similarity between the two diagrams for S_D (also not shown) because both were considerably influenced by the pattern of α-values. The overall differences between the two procedures are summarized statistically in Table II, which shows that the two sets of S_T values and the two sets of S_D values were correlated to a similar extent while to two sets of S_S values were much more poorly correlated. The mean differences between the sets of S_T, S_S and S_D values were all highly significant ($p < 0.001$), being particularly large for S_S where the largest proportion of the solute was to be found. In short, running the model with the parameters interpolated from the individual points definitely did not give the same results as running the model for the original points and interpolating the output. The reason is presumably the same as that for the effect of the unstructured variability on the functioning of the model, that is, the non-linearity of the model.

Table II: Statistical comparison of the values of S_T, S_S and S_D obtained by the two mapping approaches. Correlation and Mean difference.

	Correlation coefficient	Mean difference	SE of MD
ST	0.701	4.07	0.439
SS	0.369	-11.28	1.086
SD	0.710	7.21	0.694

6. Discussion

A computer model is an extended hypothesis. As such it should be judged, according to a dictum of Isaac Asimov, by its utility. The SLIM model is useful because it is simple and therefore fairly robust with respect to variability in its parameters. It is also useful because its parameters can be estimated from information that is readily accessible and independent of the data to be simulated. This information can be detailed, e.g. the soil moisture characteristic or the particle size distribution in top-soil and subsoil. Or, it can be at a lower level; a "menu-driven" version of the model can make use of the type of soil in the top- and sub-soils - often all the information that is available. Finally, it is useful because it has been tested against field leaching data and found to give satisfactory simulations in several different soils including one "soil" that was mainly chalk from about 0.3 m downwards. It has been put to practical use in an advisory capacity, providing information to farmers and their advisors through a viewdata system.

It is not a "research" model in that it is not fully mechanistic, but it has proved useful in a research context in two ways: in helping to partition the losses of ^{15}N-labelled nitrogen fertilizer between leaching and denitrification, and in helping to elucidate problems that arose in the use of soil mineral nitrogen as a predictor of fertilizer requirement. A precursor of the model proved valuable for intepolating the irregular breakthrough curve of ^{36}Cl-labelled chloride in the Rothamsted Drain Gauges and thereby providing a clear demonstration for the existence of mobile and immobile water in a field soil (Addiscott *et al.*, 1978).

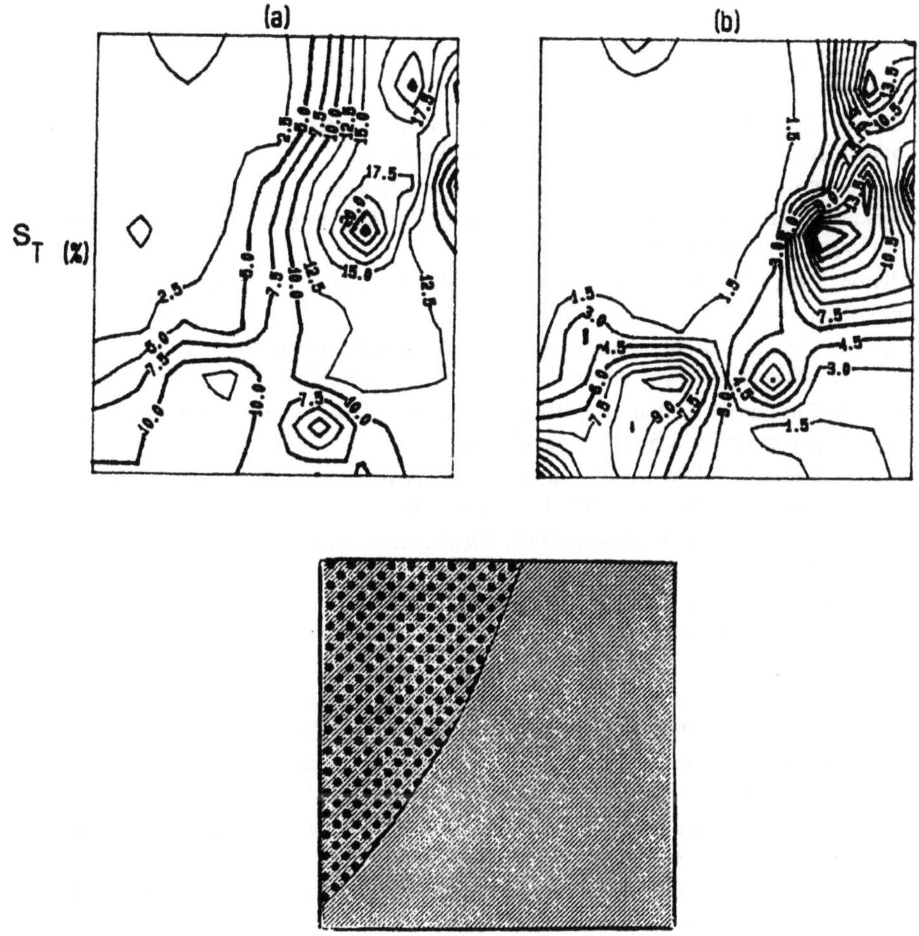

Figure 3: Contour diagrams for S_T compared with excerpt from soil survey map. (a) model output interpolated, (b) parameters interpolated, (c) soil survey map: ▨ eroded, ▩ uneroded

References

Addiscott, T.M., 1977: *A simple computer model for leaching in structured soils*. J. Soil Sci., **28**, 554-563.

Addiscott, T.M., 1982: *Simulating diffusion within soil aggregates: a simple model for cubic and other regularly-shaped aggregates*. J. Soil Sci., **33**, 37-45.

Addiscott, T.M., and G.J. Bland, 1988: In: Nitrogen efficiency in agricultural soils (Eds D.S. Jenkinson and K.A. Smith) pp 394-408. *Nitrate leaching models and soil heterogeneity*. Barking: Elsevier Applied Science.

Addiscott, T.M., P.J. Heys, and A.P. Whitmore, 1986: *Application of simple leaching models in heterogeneous soils*. Geoderma, **38**, 185-194.

Addiscott, T.M., D.A. Rose, and J. Bolton, 1978: *Chloride leaching in the Rothamsted drain gauges: influence of rainfall pattern and soil structure*. J. Soil Sci., **29**, 305-314.

Addiscott, T.M., and R.J. Wagenet, 1985a: *Concepts of solute leaching in soils: A review of modelling approaches*. J. Soil Sic., **36**, 411-424.

Addiscott, T.M., and R.J. Wagenet, 1985b: *A simple method for combining soil properties that show variability*. Soil Sci. Soc. Am. J., **49**, 1365-1369.

Addiscott, T.M., and A.P. Whitmore, 1991: *Simulation of solute leaching in soils of differing permeabilities*. Soil Use Mgmt (in preparation).

Amoozegar-Fard, A., D.R. Nielsen, and A. Warrick, 1982: *Soil solute concentration distributions for spatially-varying pore water velocities and apparent diffusion coefficients*. Soil Sci. Soc. Amer. J., **46**, 3-9.

Anderson, J.L., and J. Bouma, 1973: *Relationships between saturated hydraulic conductivity and morphometric data of an argillic horizon*. Soil Sci. Soc. Am. Proc., **378**, 408-413.

Avery, B.W. and P. Bullock, 1969: *Morphology and classification of Broadbalk soils*. Rep. Rothamsted Exp. Stn. for 1968, Pt.II., 63-81.

Burns, I.G., 1974: *A simple model for predicting the redistribution of salts applied to fallow soils after excess rainfall or evaporation*. J. Soil Sci., **25**, 165-178.

Cameron, K.C., and A. Wild, 1982: *Prediction of solute leaching under field conditions: An appraisal of three methods*. J. Soil Sci., **33**, 659-669.

Dagan, G., and E. Bresler, 1979: *Solute dispersion in unsaturated heterogeneous soil at field scale. I. Theory*. Soil Sci. Soc. Amer. J., **43**, 461-466.

Derr, B.D., R.P. Mateleski, and G.W. Peterson, 1969: *Soil factors influencing percolation test performance*. Soil Sci. Soc. Am. Proc., **33**, 942-946.

De Smedt, F., F. Wauters, and J. Sevilla, 1986: *Study of tracer movement through unsaturated sand*. Geoderma, **38**, 223-236.

Hall, D.G.M., M.J. Reeve, A.J. Thomasson, and V.F. Wright, 1977: *Water Retention, Porosity and Density of Field Soils*. Soil Survey Technical Monograph No. 9, p. 74.

Jaynes, D.B., and E.J. Tyler, 1984: *Using soil physical properties to estimate hydraulic conductivity*. Soil Sci., **138**, 298-305.

Jones, A.J., and R.J. Wagenet, 1984: *In-situ estimation of hydraulic conductivity using simplified methods*. Water Resour. Res., **20**, 1620-1626.

Jury, W.A., 1982: *Simulation of solute transport using a transfer function model*. Water Resour. Res., **18**, 363-368.

Jury, W.A., L.A. Stolzy, and P. Shouse, 1982: *A field test of the transfer function model for predicting solute transport*. Water Resour. Res., **18**, 369-375.

Lawes, J.B., J.H. Gilbert, and R. Warington, 1882: *On the amount and composition of the rain and drainage waters collected at Rothamsted. III. The drainage water from land cropped and manured*. J. Roy. Agric. Soc. Eng., **18** 1-71.

Mulla, D.J., and M.W. Hammond, 1988: In: Proceedings of the 39th Far West Regional Fertilizer Conference, pp. 169-176. *Mapping of soil test results from large irrigation circles*.

Nicholls, P.H., A. Walker, and R.J. Baker, 1983: *Measurement and simulation of the movement and degradation of atrazine and metribuzin in a fallow soil*. Pesticide Sci., **13**, 484-494.

Passioura, J.B., 1971: *Hydrodynamic dispersion in aggregated media. I. theory*. Soil Sci., **111**, 339-344.

Rao, P.S.C., P.V. Rao, and J.M. Davidson, 1977: *Estimation of the spatial variability of soil water-flux*. Soil Sci. Soc. Am. J., **41**, 1208-1209.

Ross, G.J.S., 1980: *Maximum Likelihood program*. Harpenden: Rothamsted Experimental Station.

Van Genuchten, M.Th., and P.J. Wierenga, 1976: *Mass transfer studies in porous sorbing media. I Analytical solutions*. Soil Sci. Soc. Am. J., **40**, 473-480.

Veseth, R., 1989: *Variable fertilizer application improves profits and conservation*. STEEP Extension Conservation Farming Update, Spring 1989.

Winneburger, J.T., 1974: *Correlation of three techniques for determining soil permeability*. J. Environ. Health, **37**, 108-118.

T.M. Addiscott and N.J. Bailey, IACR Rothamsted Experimental Station, Harpenden, Herts, AL5 2JQ, U.K.

Field-Scale Water and
Solute Flux in Soils
Monte Verità

PREDICTION OF CATION TRANSPORT IN SOILS USING CATION EXCHANGE REACTIONS

H. M. Selim, R. S. Mansell, L.A. Gaston, H. Flühler, and R. Schulin

Mathematical models that describe the transport of multiple ions in the soil profile are discussed. The transport models are of the convective- dispersive type and it is assumed that ion exchange is the retention reaction which governs the distribution of cations between solution and adsorbed phases. A commonly used model assumes that equilibrium reaction between solution and exchanger phases for any two cations is not influenced by other species present in solution. The necessary inputs are the cation exchange capacity and exchange selectivity coefficients for each cation pair. This classical approach has been modified by incorporating selectivity coefficients that vary with the fractional coverage on the exchanger phase. This approach more accurately represents the cation exchange reactions and has yielded improved description of Ca, Mg, and Na breakthroughs in effluent from soil columns. Nonequilibrium ion exchange behavior has been incorporated into the classical transport model using two different approaches. The first approach uses the two-region concept (mobile and immobile water) where it is postulated that physical nonequilibrium governs the ion transfer between mobile and immobile regions. The second nonequilibrium approach is based on the assumption that cation exchange reactions exhibit kinetic behavior. Depending on the type of clay and ions present in the soil, two types of cation exchange sites have been characterized; one which exhibits equilibrium and the other which shows kinetic behavior. Both nonequilibrium approaches require several additional input parameters, some of which are not easily attainable. The two-region approach has been successfully tested using several data sets from several transport studies, however, rigorous validation of the kinetic approach has not been performed. Other models requiring validation are those which account for the formation of complexes as well as ion exchange.

1. Introduction

The transport and interactions of dissolved chemicals in the soil profile are important processes which influence groundwater quality and nutrient availability in the soil root zone. The ability to describe the mobility of ion species present in the soil solution is a prerequisite to responsible use of agricultural chemicals and land disposal of chemical wastes.

Modeling cation retention and transport in the soil profile requires knowledge of several chemical and physical properties of the soil matrix, including the cation exchange capacity and the distribution of exchangeable cations between solution and sorbed phases. The work of Rubin and James (1973) is considered one of the earliest classical studies describing competitive ion exchange during the transport of multiple cation species in soils. They developed a transport model (convective-dispersive) which dealt with retention reactions of several cations during steady water flow in soils. Reversible ion exchange reactions were considered to govern the retention of cations present in the soil. In addition, the ion exchange reactions were assumed to be rapid or instantaneous which implies that local equilibrium conditions are dominant. A set of recursion equations were developed (Rubin and James, 1973) which described multispecies, heterovalent cation exchange. These equations were incorporated into the convective-dispersive equations which were then solved using a finite element approach.

Later, Valocchi et al. (1981) extended the work of Rubin and James (1973) to include multiple ion transport under conditions of varying total concentration (or ionic strength) of the soil solution. Their model was used to simulate the simultaneous transport of Na, Ca and Mg ions in a shallow underground aquifer. The analysis was applied to a field project involving direct injection of treated municipal effluent into an aquifer. Constant values of selectivity coefficients corresponding to the ionic composition of the applied effluent provided adequate prediction of effluent data. In addition, the use of ion concentrations rather than activities was shown not to restrict the predictive capability of the model.

Transport models which deal with multiple ion retention in soils include those of Miller and Benson (1983) and Cederberg et al. (1985), among others. Miller and Benson (1983) incorporated the formation of complexes in the aqueous phase and dissociation of water as well as ion exchange reactions into the transport (convective-dispersive) equation for one-dimensional flow in saturated porous media. Cederberg et al. (1985) developed a two-dimensional model for multiple ion transport. The types of chemical reactions considered were similar to those treated by Miller and Benson (1983). In addition, Cederberg et al. (1985) utilized a simplified approach in solving the system of transport and chemical equilibrium equations. Both models were used successfully to describe the Ca and Mg concentration data presented earlier by Valocchi et al. (1981).

In an attempt to describe the transport and exchange reactions of cations in aggregated porous media, the physical nonequilibrium approach of the mobile-immobile (or two-region) concept was utilized. van Eijkeren and Loch (1984), Schulin et al. (1986), and Mansell et al. (1986) considered the exchange of cations to be governed by a general form of the exchange equation which was then incorporated into the mobile-immobile, convective-dispersive transport equation. The capability of this approach was examined by Selim et al. (1987) for describing the mobility of Ca and Mg ions in soil columns, packed with 1-2 and 2-4 mm aggregates, under conditions of constant and variable ionic strength of the soil solution. Recently, Mansell et al. (1988) examined this mobile-immobile approach for the transport and exchange reactions for ternary systems (Na-Ca-Mg) in soil. In their analysis, Mansell et al. (1988) allowed cation exchange selectivities to vary with fractional coverage of the exchange sites. They found that incorporation of variable cation exchange selectivities and the immobile-immobile approach provided an improved description of cation breakthrough results.

In this paper, an overview of ion retention and transport models is presented. Jury et al. (1988) have discussed in detail a number of difficulties with experimental validation of solute transport models at the larger field scale. Steady one-dimensional water flow through a uniform, homogeneous, isotropic soil is assumed here. Although these models can be used to describe solute transport in a soil profile under simplified flow conditions in the field, the primary intent here is to emphasize the importance of ion exchange as the chemical retention mechanism which governs the distribution of multiple cations between solution and adsorbed phases during water flow in soil. Affinity of cations for exchange sites in the soil matrix is described by exchange selectivity coefficients for ion species present in the soil-water system. Emphasis is on chemical models where nonequilibrium ion exchange behavior is described using physical (two-region) or chemical (two-site) kinetic approaches. Input data required for modeling management systems on a field scale are also identified.

2. Ion Exchange Models

2.1 Classical Approach

The classical convective-dispersive equation which is generally acceptable for describing the transport of solutes in soils in one dimension may be expressed as

$$\theta \frac{\partial C}{\partial t} + \rho \frac{\partial S}{\partial t} = \theta D \frac{\partial^2 C}{\partial x^2} - v \frac{\partial C}{\partial x} \tag{1}$$

where C is solute concentration in soil solution (mmol(+)/ml), θ is the soil water content (cm^3/cm^3), ρ is the soil bulk density (g/cm^3), D is the hydrodynamic dispersion coefficient (cm^2/day), and v is Darcy's water flux density (cm/day). Here, S denotes the solute concentration associated with the solid phase of the soil (mmol(+)/g soil) and the term ($\partial S/\partial t$) represents a fully reversible reaction between the solution and the solid matrix.

Furthermore, we consider ion exchange as the governing mechanism for the retention of cations on soil matrix surfaces. In a generalized form, ion exchange reactions for two ions i and j may be written as (Sposito, 1981)

$$K_{ij} = \frac{\left(s_i/c_i\right)^{v_j}}{\left(s_j/c_j\right)^{v_i}} \tag{2}$$

where K_{ij} is the selectivity coefficient of ions i over j. In addition, c_i and c_j are the relative ion concentrations (dimensionless) such that $c_i = C_i/C_T$ and $c_j = C_j/C_T$ where C_i and C_j (mmol(+)/ml) are the concentrations in the soil solution of ions i and j, and C_T is the total concentration (mmol(+)/ml). Also s_i and s_j are amounts retained on the solid matrix surfaces (dimensionless) and are expressed as equivalent fractions where $s_i = S_i/S_T$ and $s_j = S_j/S_T$. Here, S_i and S_j are the amounts adsorbed (mmol(+)/g soil) and S_T is the cation exchange (or adsorption) capacity of the soil (mmol(+)/g soil). For simplicity, we consider the case of binary exchange of monovalent ions; i.e., $v_1 = v_2 = 1$. Therefore, we have the relation (Valocchi et al., 1981)

$$K_{12} = \frac{s_1/c_1}{s_2/c_2} \tag{3}$$

Assuming that the total solution concentration is time invariant, $C_T = C_1 + C_2$ = constant. In addition, we consider the cation exchange capacity an intrinsic property of the soil, i.e., S_T is constant (= $S_1 + S_2$). Following the work of Rubin and James (1973) and Valocchi et al. (1981), eq. (3), upon rearrangement, yields the following isotherm for the adsorbed phase equivalent fraction (s_1) versus concentration (c_1),

$$s_1 = \frac{K_{12}\, c_1}{1+(K_{12}-1)\, c_1} \tag{4}$$

Since C_T and S_T are assumed constant the respective values for ion 2 (i.e., c_2 and s_2) can be easily obtained. This isotherm relation eq. (4) indicates that for $K_{12} \neq 1$, we have a nonlinear type sorption isotherm. Upon incorporation of the above ion-exchange sorption isotherm into the transport eq. (1), we have

$$R\frac{\partial c_1}{\partial t} = D\frac{\partial^2 c_1}{\partial x^2} - \frac{v}{\theta}\frac{\partial c_1}{\partial x} \tag{5}$$

where the term R may be referred to as the retardation factor expressed as,

$$R = 1 + \frac{\rho S_T}{\theta C_T\left(1+(K_{12}-1)c_1\right)^2} \underline{K_{12}} \tag{6}$$

and R is a function of concentration for $K_{12} \neq 1$.

Breakthrough results for Ca and Mg leaching from Abist soil columns, packed uniformly with two aggregate sizes, are shown in Figure 1. These results are from Selim et al. (1987) where a Mg pulse was introduced into Ca-saturated soil columns and the total concentration (C_T) was maintained constant. The solid and dashed lines shown are model calculations of Ca and Mg results based on the classical convective-dispersive transport equation with ion exchange as the retention mechanism. In general, adequate model predictions for both ions were achieved for the two aggregate sizes for early times or pore volumes. Less than adequate predictions were obtained, however, for large times, i.e., during leaching of the Mg pulse by Ca. Selim et al. (1987) suggested that such model deviations may be due to lack of complete local equilibrium between the ions in solution and those on the ion exchange surfaces.

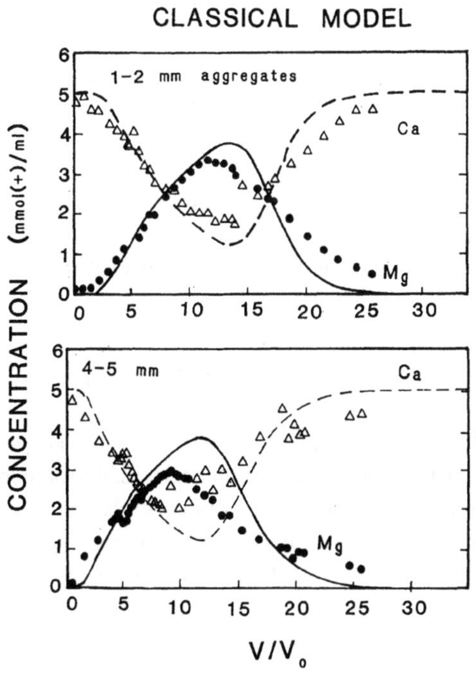

Figure 1: Calcium and Mg breakthrough results from soil columns for two aggregate sizes of Abist soil. Predictions obtained using the classical model are shown by the smooth curves.

2.2 *Mobile-Immobile Approach*

The mobile-immobile or two-region concept is based on the assumption that the soil water content θ can be divided into two parts, an immobile water content and a mobile water content. Specifically, we assume θ^m (cm^3/cm^3) as the mobile water content which is present inside large (interaggregate) pores where solute transport occurs by convection and dispersion. The immobile water (θ^{im}) is located inside aggregate pores (intra-aggregate) where the solute transfer occurs by diffusion only. Moreover, C_i^m and C_i^{im} are the concentrations of ion i in the mobile and immobile phases, respectively. We further assume that the soil matrix is made up of two regions (or sites), a dynamic or easily accessible soil region and a less accessible soil region (for details see van Genuchten and Wierenga, 1976). Furthermore, S_i^m and S_i^{im} represent the amounts of ion i retained on the matrix surfaces of the dynamic and the less accessible regions, respectively.

The general form of the transport equation for the two region (mobile-immobile) concept can be written (for ion i) as

$$\theta^{im}\frac{\partial C_i^{im}}{\partial t}+\rho\,(1\text{-}f)\frac{\partial S_i^{im}}{\partial t}+\theta^{m}\frac{\partial C_i^{m}}{\partial t}+\rho f\frac{\partial S_i^{m}}{\partial t}= \tag{7}$$

$$=\theta^{m}D^{m}\frac{\partial^2 c_i^{m}}{\partial x^2}-v\frac{\partial C_i^{m}}{\partial x}$$

where f represents the ratio of dynamic or active sites to the total sites. The transfer equation governing the interaction between the mobile and immobile phases (for ion i) is

$$\theta^{im}\frac{\partial C_i^{im}}{\partial t}+\rho\,(1\text{-}f)\frac{\partial S_i^{im}}{\partial t}=\alpha\left(C_i^{m}-C_i^{im}\right) \tag{8}$$

where α is the mass transfer coefficient (day^{-1}) between the mobile and immobile phases. To describe ion retention in the dynamic and less accessible regions, the sorption terms in the above two equations may be expressed using the equilibrium ion exchange relationship of eq. (3). A common assumption is that such a relationship is valid for the exchange sites for both regions. Specifically, the ion selectivity coefficient K_{12} for the dynamic region is considered the same as that for the less accessible sites, i.e.,

$$\left(K_{12}\right)^{m}=\left(K_{12}\right)^{im}=K_{12} \tag{9}$$

Such an assumption was used by van Eijkeren and Loch (1984) and Selim et al. (1987) for ion exchange reactions.

In order to test the capability of the mobile-immobile concept of describing the transport of cations, the breakthrough results shown previously (see Figure 1) were utilized. Model predictions of Ca and Mg concentrations in the leachate for the two aggregate sizes are given by the solid and dashed lines shown in Figure 2. Obvious improvements in model predictions, in comparison to the classical approach (see Figure 1), were achieved. These improvements were obtained for large pore volumes following the application of the Mg pulse. Improvements in predictions using the mobile-immobile model may be considered as evidence of lack of complete local equilibrium between the ions in the soil solution and those on the exchange surfaces. Improved predictions using the two-region approach were achieved by Schulin et al. (1989) for Ca and Mg transport under conditions of variable ionic strength for two well-aggregated forest soils. Mansell et al. (1988) also reported improved predictions using the two-region approach for a ternary (Na-Ca-Mg) system in a Yolo loam soil.

2.3 Variable Selectivities

In the development of the classical and mobile-immobile concepts to describe the transport and exchange of ions in soils, several simplifying assumptions were made. These include constant cation exchange capacity (CEC); ion activities in solution and on exchange surfaces can be approximated by their concentrations; and the validity of the local equilibrium assumption for ion exchange reactions. A common assumption is that the distribution of each pair of cations, i and j, can be described by a constant selectivity coefficient K_{ij}. Such assumptions make it possible to arrive at recursion formulas for multiple ions, as were introduced by Rubin and James (1973). Generalized isotherms for multiple ions were based on binary exchange coefficients for all combinations of ions present in the soil system. Such generalized isotherms were recently used by Valocchi et al. (1981) and Mansell et al. (1986). As pointed out by Mansell et al. (1988), however, the assumption of constant exchange selectivity is unfounded as evidenced by the majority of isotherm data in the literature (Sposito, 1981; Jardine and Sparks, 1984; Parker and Jardine, 1986). As a result, K_{ij} coefficients are no longer constant but vary with the relative fraction of cations on the exchange surfaces.

TWO-REGION MODEL

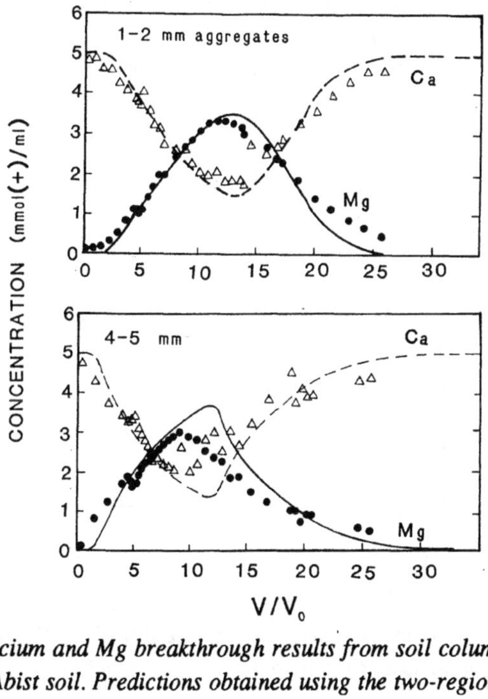

Figure 2: Calcium and Mg breakthrough results from soil columns for two aggregate sizes of Abist soil. Predictions obtained using the two-region model are shown by the smooth curves

Mansell et al. (1988) utilized the sorption isotherm results of Lai et al. (1978) for Mg-Ca and Na-Ca in order to calculate binary selectivity coefficients K_{ij} versus relative ion concentration

(C/C_T) shown in Figure 3. The dashed curves represent least-squares best-fit of the data to the empirical relation $\{\log(K_{ij}) = a + bC\}$. The results clearly show a strong K_{ij} dependency over the concentration range with high affinity of Mg over Ca for C/C_T less than 0.4. As expected, adsorption of Ca or Mg was preferred to adsorption of Na. However, the results show an increased K_{ij} for Na→Ca at low (C/C_T <0.4) and high (C/C_T > 0.85) relative concentrations. The influence of variable selectivity coefficients on the predictions of cation transport was investigated by Mansell et al. (1988). The transport data sets used were those from miscible displacement experiments of Lai et al. (1978). Mansell et al. (1988) incorporated additional terms into the classical (convective-dispersive) equation in order to account for variable K_{ij}'s for the ternary systems (Na-Ca-Mg) of Lai et al. (1978). They also utilized the mobile-immobile approach in conjunction with variable K_{ij}'s in order to describe the same data set. Mansell et al. (1988) found that good breakthrough predictions were obtained for the relatively noncompetitive Na when either constant or variable K_{ij}'s were used with the classical model (see Figure 4). However, the use of constant K_{ij}'s underestimated the tailing of the Mg breakthrough data. Description of Mg tailing was improved when variable K_{ij}'s were used but the extent of Mg retardation (peak location) was somewhat overestimated. The combined use of the mobile-immobile approach and variable K_{ij}'s provided the best overall description of Na and Mg breakthrough results.

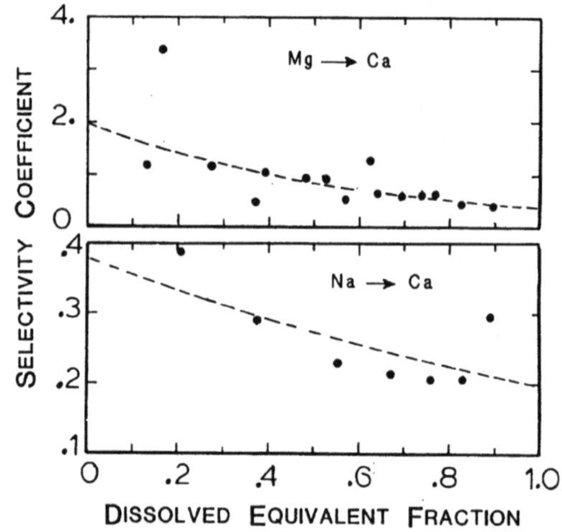

Figure 3: Exchange selectivity coefficient K_{ij} versus equivalent fraction of dissolved species (C/C_T) for Mg-Ca and Na-Ca for Yolo soil (Lai et al., 1978). Smooth curves were obtained using least-square fitting to the equation $\{\log(K_{ij}) = a + bC\}$.

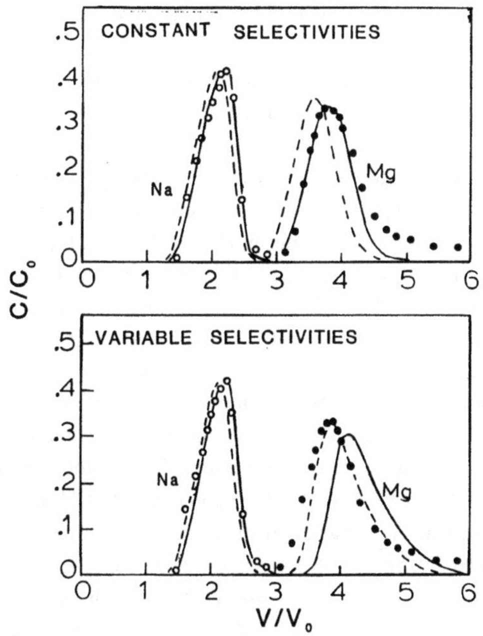

Figure 4: Sodium and Mg breakthrough results from soil columns of Yolo soil (Lai et al., 1978). Solid and dashed curves are predictions using the classical and two-region model, respectively. Exchange selectivity coefficients used in model predictions were considered constant (top figure) or variable (bottom figure).

2.4 Complex Formation

The approaches described above deal only with uncomplexed cations and their associated exchange reactions. Recently, several models that include formation of complexes along with ion exchange have been introduced. These models represent an extension of the classical approach where multiple ion transport and reactions are considered. Examples of such models include those of Miller and Benson (1983) and Cederberg et al. (1985). These models were utilized to investigate the influence of complexation on transport predictions of cations. However, rigorous validation of such models using experimental data sets based on carefully planned studies is still needed.

In an attempt to investigate the contribution of formation of complexes on the predictive capability of ion exchange models on cation transport, Gaston and Selim (1990) carried out batch and miscible displacement experiments on a Sharkey clay soil. Binary (Ca-Mg) and ternary (Ca-Mg-

Na) exchange systems, with Cl as the counter ion, were studied. Both Ca^{2+} and Mg^{2+} form complexes with Cl^-. Sposito et al. (1983a and 1983b) have inferred that these complexes adsorb onto montmorillonite clay.

Gaston and Selim (1990) extended the classical transport model to include $CaCl^+$ and $MgCl^+$. The distribution of these complexes between solution and sorbed phases was modeled as a cation exchange process. Cation exchange selectivities involving these complexes were estimated from Ca-Mg, Ca-Na, and Mg-Na exchange data. Despite good description of exchange isotherm results without consideration of complex formation, some uncertainty remained as to whether explicit inclusion of complexes could offer better model predictions. Figure 5 shows transport data for the miscible displacement of a pulse of Mg in Ca-saturated soil columns. Two pore water flow velocities (v/θ) were used. The solid curves represent calculations using the classical transport model where complexation was ignored. The expanded model included complex formation and adsorption of complexes. As well as transport of the basic cations, transport of free and complexed Cl^- were modeled. The model was formulated so as to calculate total Ca, Mg, and Cl according to the tenad concept of Rubin (1983). Figure 6 shows model predictions for total Ca and Mg elution (Ca-Mg exchange), for the lower and higher velocity experiments, respectively, compared to experimental data. Inclusion of the complexes $CaCl^+$ and $MgCl^+$ offered no improvement over treating all Ca and Mg as divalent species. In fact, partitioning these species between divalent and complexed forms resulted in somewhat greater predicted retardation of Mg elution. This result reflects all errors attendant to estimating the suite of selectivity coefficients involving $CaCl^+$ and $MgCl^+$. Similar results were obtained for Mg leaching in the Ca-Mg-Na system. Adequate description of Na elution was obtained using either the classical or the extended model.

2.5 *Kinetic Approach*

In the above approaches, ion exchange reactions were assumed instantaneous in nature. Such an assumption of local equilibrium has been supported by numerous investigations for a wide range of cations and exchange materials. However, several studies showed cation exchange to be a kinetic process in which local equilibrium was not reached before several days of contact time. Sparks (1989) compiled an extensive list of cations (and anions) that exhibited kinetic ion exchange behavior in soils. Examples of such ions include Al, NH_4, K and several heavy metal cations. According to Ogwada and Sparks (1986), observed kinetic ion exchange behavior in soils is probably due to mass transfer (or diffusion) and chemical kinetic processes. They stated that for chemical sorption to occur, ions must be transported to the active (fixed) sites of the soil particles. The film of water adhering to and surrounding the particles and water within the interlayer spaces of the particles are both zones of low concentrations due to depletion by adsorption of ions onto the exchange sites. The decrease in concentration in these two interface zones is compensated by diffusion of ions from the bulk solution.

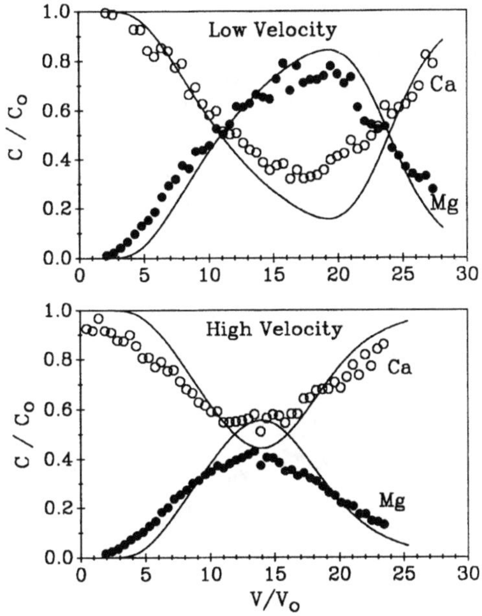

Figure 5: Calcium and Mg breakthrough results from soil columns for two water flow velocities (0.296 and 0.595 m day) of Sharkey clay soil. Predictions obtained using the classical model are shown by the solid curves.

Types of clay minerals which dominate a particular soil influence the extent of kinetic behavior of ion exchange reactions. Several studies indicate that ion exchange on kaolinite, smectite, and illite are rapid and local equilibrium is often observed within a few seconds. In contrast, ion exchange on vermiculite and micaceous minerals, particularly involving ions such as K, NH_4, and Ce, are strongly time dependent. It is postulated that in such 2:1 type minerals, intra-particle diffusion is a rate-controlling mechanism governing the kinetics of adsorption of cations (for a review see Sparks, 1989). The work of Jardine and Sparks (1984) is one of several studies which illustrate that rates of potassium sorption-desorption (in K-Ca system) can be rapid for kaolinite and montmorillonite. However, the rate of potassium exchange was slow for vermiculite. In addition, Jardine and Sparks (1984) showed that kinetics for K as determined using batch type experiments were well described using a two-site approach of the first-order kinetic type. They attributed this behavior to the presence of exchange sites of varying reactivity (and/or accessibility) to K and Ca ions.

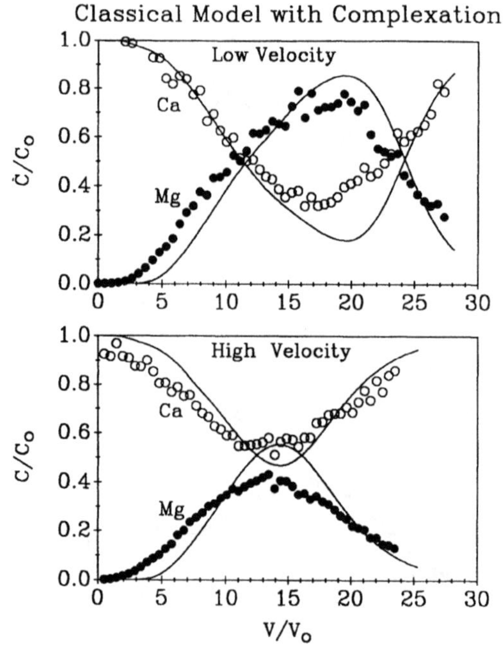

Figure 6: Calcium and Mg breakthrough results from soil columns for two water flow velocities (0.296 and 0.595 m/day) of Sharkey clay soil. Smooth cures are predictions based on the classical model when complexation reactions were incorporated.

Parker and Jardine (1986) extended the work of Jardine and Sparks (1984) by incorporating ion exchange reactions with the two-site approach. The model considers two types of cation exchange sites, one which exhibits local equilibrium and the other which shows kinetic behavior. The proposed kinetic ion-exchange reaction was analogous to mass transfer or diffusion between the solid and solution phase,

$$\frac{\partial s_e}{\partial t} = \omega\left(s_e^* - s_e\right) \tag{10}$$

where s_e is the amount sorbed on matrix surfaces, s_e^* is the equilibrium sorbed amount (at time t) and ω is an apparent rate coefficient (day^{-1}) for the kinetic-type sites. Expressions similar to the above have been used to describe mass transfer between mobile and immobile water as well as chemical kinetics.

Recently, we obtained apparent rate coefficients (ω), associated with the kinetic sites, using best-fit of batch (desorption) kinetic data for Ca desorption from a Cecil (surface and subsurface) soil. Furthermore, we tested the capability of the kinetic ion exchange approach in describing transport data obtained from miscible displacement experiments. Figure 7 shows effluent Ca^{2+} concentrations for the surface Cecil soil leached with pH 3.0 H_2SO_4 and HNO_3 solutions. These solutions were introduced into columns packed with air-dry soils (Gaston et al., 1987). Model predictions given in Figure 7 (top) were based on an assumption of equilibrium cation exchange and are shown as smooth curves. Although the model distinguished differences in Ca^{2+} breakthroughs which depended on the type of input anion, predicted breakthroughs lagged the experimental results and peak concentrations were overestimated. In addition, the equilibrium model did not predict the observed tailing of the measured breakthrough results. Use of a kinetic-equilibrium (two-site) ion exchange approach resulted in a significant improvement of the prediction capability of the breakthrough result (see Figure 7). Observed tailing in the effluent results was well described using the kinetic model. Although concentration maximum for Ca^{2+} leached by H_2SO_4 was overestimated, model calculations provided adequate overall predictions of the breakthrough data. Discrepancies between predicted and observed leaching behavior reflect, in part, the simplified model used to calculate transport and reaction under conditions of transient water flow. Despite such simplifications, the kinetic approach provided good descriptions of peak locations and introduced tailing.

3. Summary and Conclusions

An overview of selected transport models was presented which included ion exchange as a retention process governing the distribution of cations between solution and adsorbed cations. Use of ion exchange reactions offers several advantages over other approaches. Unlike empirical approaches such as the use of Freundlich or Langmuir isotherms where only one of several species is considered, the ion exchange concept is based upon the competitive reactions of two or more species simultaneously. In the classical approach considered above, selectivity coefficients of the ions present in the soil system were assumed invariant with concentration on the exchanger phase. The selectivity coefficient represents a measure of the degree of affinity of ions for exchange sites on matrix surfaces. The classical approach provided adequate description of cation mobility in binary systems for several data sets. The use of variable selectivity coefficients as a function of the fraction of ions on the exchanger phase proved superior in describing ternary ion systems in comparison to the classical approach when constant selectivities were assumed.

Ion exchange models which account for nonequilibrium behavior of cation during transport were presented. Two nonequilibrium type approaches were discussed. The first approach was the two-region type in which physical nonequilibrium is assumed to govern ion transfer between mobile and immobile regions. The second nonequilibrium approach was based on kinetics of ion exchange. Both nonequilibrium approaches require several additional input parameters, some of which are not easily measured. The two-region approach has been successfully used to account

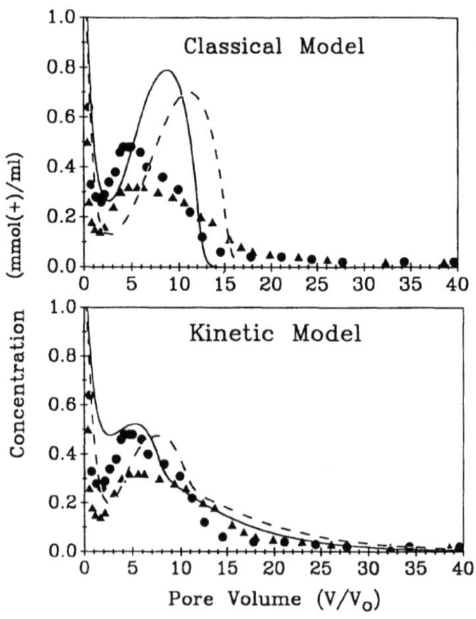

Figure 7: Calcium breakthrough results from soil columns of Cecil soil which received continuous application of pH 3.0 H_2SO_4 solution (closed circles) or HNO_3 solution (triangles). Solid and dashed curves are predictions using the classical model (top figure) or two-site, equilibrium-kinetic model (bottom figure). Solide and dashed curves are predictions corresponding to the H_2SO_4 and HNO_3 input solutions, respectively.

for the apparent lack of equilibrium as manifested by excessive tailing of breakthrough results observed in several data sets. In contrast, additional validation of the kinetic approach is needed. In conclusion, incorporation of ion exchange submodels into transport models represent powerful tools for the prediction of the fate of cations in the soil profile. An important feature inherent to such models is that they account for all major cationic species in soil solution. In fact, models which incorporate the formation of complexes and dissociation of water along with ion exchange deal with cationic as well as anionic species. However, a major disadvantage of such models is that all of the required soil parameters are rarely measured on a routine basis. Moreover, owing to the spatial (vertical and lateral) and temporal variability of soil properties, use of single values, even field scale averages, for model parameters may lead to erroneous predictions of field scale chemical transport. Attempts to account for such variability in modeling transport have been based on the convective-dispersive model (eq. 1). In such application, eq. (1) assumes the character of a stochastic equation (Sposito et al., 1986). Alternatively, field scale solute transport may be described independently of any particular mechanisms if a transfer function model is used (Jury et al., 1986). The later models, however, do not lend themselves easily to systems of competitive ions where the equilibrium/kinetic reactions are often nonlinear.

Acknowledgements

This work was supported in part by the U.S. National Science Foundation (grant no. INT-8713283) and the Swiss National Science Foundation and is published as Manuscript no. 90-09-4192 of the Louisiana Agricultural Experiment Station, LSU Agricultural Center, Baton Rouge, Louisiana 70803.

References

Cederberg, G. A., R. L. Street, and J. O. Leckie, 1985: *A groundwater mass transport and equilibrium chemistry model for multicomponent systems*. Water Resour. Res., **21**,1095-1104.

Gaston, L. A. and H. M. Selim. 1990: *Transport of exchangeable cations in an aggregated clay soil.* Soil Sci. Soc. Am. J., **54**,31-38.

Gaston, L. A., R. S. Mansell, R. D. Rhue, S. A. Bloom, and B. G. Volk. 1987: *Cation leaching during application of sulfuric and nitric acid to an Ultisol.* In R. Perry (ed) Acid Rain: Scientific and Technical Advances. Selper, Ltd., London (pp 421-428).

Jardine, P.M., and D.L. Sparks, 1984:.*Potassium-calcium exchange in a multireactive soil system, I. Kinetics*. Soil Sci. Soc. Am. J., **48**, 39-45.

Jury, W. A., G. Sposito, and R. E. White. 1986: *A transfer function model of solute transport through soil; 1. Fundamental concepts.* Water Resour. Res., **22**, 243-247.

Jury, W. A., G. L. Butters, L. D. Clendening, F. F. Ernst, H. Elabd, T. R. Ellsworth, and L. H. Stolzy. 1988: *Validation of solute transport models at the field scale. Proc. Int. Conf. and Workshop on Validation of Flow and Transport Models for the Unsaturated Zone, Ruidoso.* New Mexico, May 22-25, 1988, P. J. Wierenga and D. Bachelet (eds.), pp. 206-218, New Mexico State Univ., Las Cruces, 545 p.

Lai, Sung-Ho, J. J. Jurinak, and R. J. Wagenet, 1978: *Multicomponent cation adsorption during convective-dispersive flow through soils, Experimental study.* Soil Sci. Soc. Am. J., **42**, 240-243.

Mansell, R.S., S.A. Bloom, H.M. Selim, and R.D. Rhue, 1988: *Simulated transport of multiple cations in soil using variable selectivity coefficients.* Soil Sci. Soc. Am. J., **52**, 1533-1540.

Mansell, R.S., S.A. Bloom, H.M. Selim, and R.D. Rhue, 1986: *Multispecies cation leaching during continuous displacement of electrolyte solutions through soil columns.* Geoderma, **38**, 61-75.

Miller, C.W., and L.V. Benson, 1983: *Simulation of solute transport in a chemically reactive heterogeneous system. Model development and application.* Water Resour. Res., **19**, 381-391.

Ogwada, R. A. and D. L. Sparks, 1986: *Kinetics of ion exchange on clay minerals and soil, I. Evaluation of methods.* Soil Sci. Soc. Am. J., **50**, 1158-1162.

Parker, J. C. and P. M. Jardine, 1986: *Effect of heterogeneous adsorption behavior on ion transport.* Water Resour. Res., **22**, 1334-1340.

Rubin, J, 1983: *Transport of reacting solutes in porous media: Relation between mathematical nature of problem formulation and chemical nature of reactions.* Water Resour. Res, **19**, 1231-1252.

Rubin, J., and R.V. James, 1973: *Dispersion-affected transport of reacting solutes in saturated porous media: Galerkin method applied to equilibrium controlled exchange in unidirectional steady water flow.* Water Resour. Res., **9**, 1332-1356.

Schulin, R., H. Flühler, R.S. Mansell, and H.M. Selim, 1986: *Miscible displacement of ions in aggregated soils.* Geoderma, **38**, 311-322.

Schulin, R., A. Papritz, H. Flühler, and H.M. Selim, 1989: *Calcium and magnesium transport in aggregated soils at variable ionic strength.* Geoderma, **44**, 129-141.

Selim, H.M., R. Schulin, and H. Flühler, 1987: *Transport and ion exchange of calcium and magnesium in an aggregated soil.* Soil Sci. Soc. Am. J., **51**, 876-884.

Sparks, D. L. 1989: *Kinetics of soil chemical processes.* Academic Press, San Diego.

Sposito, G. 1981: *The thermodynamics of soil solution.* Oxford University Press, N.Y.

Sposito, G., K.M. Holtzclaw, L. Charlet, C. Jouany, and A.L. Page, 1983a: *Sodium-calcium and sodium-magnesium exchange on Wyoming bentonite in perchlorate and chloride ionic media.* Soil Sci. Soc. Am. J., **47**, 51-56.

Sposito, G., K.M. Holtzclaw, C. Jouany and L. Charlet, 1983b: *Cation selectivity in sodium-calcium, sodium-magnesium, and calcium-magnesium exchange on Wyoming bentonite at 298K.* Soil Sci. Soc. Am. J., **47**, 917-921.

Sposito, G., W. A. Jury, and V. K. Gupta. 1986: *Fundamentalist problems in the stochastic convection-dispersion model of solute transport in aquifers and field soils.* Water Resour. Res., **22**, 77-88.

van Eijkeren, J. C. M., and I. P. G. Lock. 1984: *Transport of cation solutes in sorbing porous medium.* Water Resour. Res., **20**, 714-718.

van Genuchten, M. Th., and P.J. Wierenga, 1976: *Mass transfer studies in sorbing porous media. I. Analytical solutions.* Soil Sci. Soc. Am. J., **40**, 473-480.

Valocchi, A.J., R.L. Street, and P.V. Roberts, 1981: *Transport of ion-exchange solutes in groundwater: Chromatographic theory and field simulations.* Water Resour. Res., **17**, 1517-1527.

H. M. Selim*, R. S. Mansell**, L.A. Gaston*, H. Flühler***, and R. Schulin***

*Department of Agronomy, Louisiana State University, Baton Rouge, LA 70803, USA.

**Soil Science Department, University of Florida, Gainesville, FL 32611, USA.

***Laboratory of Soil Physics, Swiss Federal Institute of Technology, Zürich, Switzerland.

Field-Scale Water and
Solute Flux in Soils
Monte Verità

TRANSPORT OF A CONSERVATIVE TRACER UNDER FIELD CONDITIONS: QUALITATIVE MODELLING WITH RANDOM WALK IN A DOUBLE POROUS MEDIUM

K. Roth, H. Flühler, and W. Attinger

We propose a simple conceptual model for the transport of a conservative tracer through a field soil under natural conditions. The soil is modelled as a medium consisting of two homogeneous porous structures described by very different transport parameters. The transport of a conservative solute through this medium is simulated by a random walk of a large number of particles that do not interact with each other. In each homogeneous structure the movement of the particles is described by a convection-dispersion equation in the vertical and by a diffusion equation in the horizontal direction. The diffusion-dispersion processes are simulated by a random walk. The results of these simulations were found to be in good qualitative agreement with concentrations that were measured in a tracer experiment.

1. Introduction

Despite many efforts to understand the transport of solutes on the field scale under natural conditions, the processes involved are not completely understood. One intriguing feature found in several experiments is the splitting of an originally single tracer pulse into two parts, moving at very different velocities (Jury et al., 1986, Roth et al., 1988). Apart from practical consequences for the groundwater quality, these experiments also reveal a fundamental deficiency in most theories concerned with solute transport in the vadose zone. Theories are usually based on the assumption that soil is a homogeneous medium. Together with some additional assumptions, this leads to the convection-dispersion model of solute transport in soils (Sposito et al., 1979). This model does normally not cause pulse splitting because the parameter fields are assumed to be constant or varying only slightly.

A different approach starts from the assumption that soil is a superposition of two interacting media, each of which is "sufficiently" represented in any physically infinitesimal volume element of the soil (Aifantis, 1979). Transport in both media is described by the convection-dispersion model with different parameters. The interaction between the two media is modelled as a linear exchange, which means that the mass flux between the media is proportional to the concentration difference. The resulting set of coupled partial differential equations has been solved semi-analytically for the case of constant parameters and was shown to generate pulse splitting (Walker, 1987).

In this paper, we develop a model that extends this approach by dropping the assumption that each medium is present in any physically infinitesimal volume. We assume that the soil is a sum of two structures described by very different transport parameters and the solute is simulated by a large number of particles that do not interact with each other. The vertical movement of the particles in each structure is described by a convection-dispersion equation and the horizontal movement by a diffusion equation. The diffusive-dispersive component of the movement is simulated by a random walk.

We will use data from an experiment where the movement of chloride in a natural soil was monitored with a high spatial and temporal resolution (Roth, 1989). In this experiment, the measured chloride concentrations showed a distinct splitting of the applied tracer pulse.

2. Tracer Experiment

2.1 Material and Methods

The experiment was carried out in a typical parabrown earth. Estimated texture of the layers down to a depth of 3 m is summarized in Table I. Below 3 m the soil consists of very impermeable clayey-loamy layers.

To facilitate measuring solute concentrations in the undisturbed soil, a tunnel, 12 m long and 3 m deep, had been constructed. From this tunnel a vertical transect 10 m long and extending between 0.4 and 2.4 m below the soil surface was instrumented with 110 ceramic suction cups. The cups were installed horizontally at a distance of about 1 m from the tunnel wall. Horizontal

Table I: Estimated texture of the different layers in the soil profile (determined by M. Müller, Forschungsanstalt für Pflanzenbau, Reckenholz).

depth [cm]	org. matter [%w]	clay [%w]	silt [%w]	sand [%w]	skeleton [%v]
0-30	5	19	35	46	1
30-80	1	22	35	43	2
80-100	1.5	21	35	44	7
100-140	0.5	26	32	42	8
140-160	0	30	32	38	30
160-220	0	19	30	51	35
220-225	0	<5	<10	>85	24
225-300	0	18	35	47	3

spacing between the instruments was 1 m and vertical spacing 0.2 m. Samples of the soil solution were taken at time intervals between one day and one week, depending on infiltration rate. Samples were taken simultaneously by lowering the air pressure in the suction cups to 400 mbar. After extracting about 10 ml of soil solution with every suction cup, pressure was restored to normal level.

The tracer was applied homogeneously as a mixture of three chloride salts (NH_4Cl, KCl, and $MgCl_2$) with an areal dose of 5.6 mole Cl m^{-2}. The area was chosen large enough to prevent any influence of the boundaries on the measured concentrations. Five days after the application, the salts had been totally dissolved by 6 mm of natural, low intensity precipitation.

The total infiltration during the whole experiment was 853 mm. During the first phase 99 mm of natural rainfall (maximal intensity: 10 mm/h) infiltrated. In the second phase 601 mm of water were applied with a sprinkler system (maximal intensity: 8 mm/h). The last 153 mm of water consisted again of natural precipitation.

The mean volumetric water content in the measuring region was measured with TDR probes. During the tracer experiment it was 0.38±0.04.

2.2 Results

For the current study, concentrations measured in the two-dimensional vertical transect are averaged horizontally resulting in a mean vertical concentration profile. This averaging gives a first order approximation to the real field using a horizontally homogeneous medium. The movement of the tracer is described as a function of cumulative infiltration as this is a more appropriate variable in non-steady situations then time (Beese and Wierenga, 1980).

The measured mean concentrations (fig. 1) show clearly the splitting of the tracer pulse. After 31 mm of natural precipitation, mean Cl-concentrations 2.2 m below ground increased by almost an order of magnitude. This increase was due to very high values measured in only a few of the suction cups. Concentrations at this depth continued to increase up to a cumulative infiltration of about 150 mm and decreased steadily afterwards. This fast transport is contrasting with the slowly moving part of the tracer in the upper layers of the measuring region. (Assuming convective-dispersive transport in a homogeneous medium with a mean water content θ=0.38, the peak concentration after the infiltration of 853 mm of water would be expected at a depth of 2.24 m.) The measured concentrations in the soil between these two pulses are increasing only very slowly and do not show any traces of the fast pulse that had to be transported through this zone.

To explain the measurements the following conception is proposed. (i) The soil is assumed to be a sum of two interacting structures described by very different transport parameters. The two structures will be called micro- and macrostructure, respectively. (ii) Sudden changes in the soil texture between 2.2 and 3.0 m below ground (Tab. I) lead to impermeable layers which increase the interaction between the two structures. (iii) The measurements of the suction cups represent the concentrations in the microstructure only. (The probability of measuring a pulse moving in the macrostructure is the product of two small numbers, namely of the probability that a suction

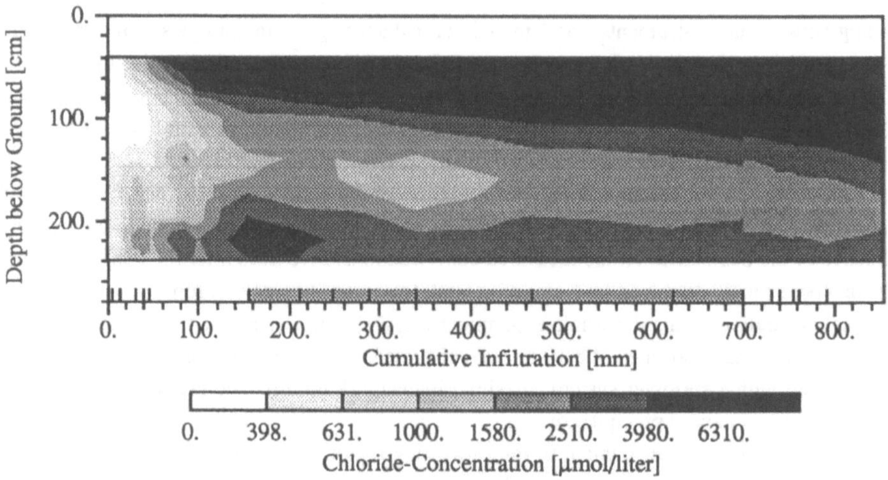

Figure 1: Mean Cl-concentration as a function of depth below ground and cumulative infiltration since beginning of tracer application. The ticks on the horizontal axis inside the frame indicate times where solution samples were taken. Concentrations between measurements are linearly interpolated. During the period marked by the gray bar the field was artificially irrigated. (Note logarithmic scale for concentrations.)

cup is installed within the macrostructure and of the probability that the sample is taken during the short time when the pulse is moving past this cup.)

3. Model

The model proposed below is intended to demonstrate that the results of the tracer experiment can be understood qualitatively with a few simplifying assumptions.

3.1 Concept

Solute transport is modelled as the movement of a large number of particles in a double porous random medium. This means, that the *transport volume* of the medium consists of the sum of two stationary stochastic structures, called micro- and macrostructure, with different transport properties. The fractions of the transport volume belonging to the micro- and macrostructure are ϑ_1 and $1-\vartheta_1$, respectively. Micro- and macrostructure can be interpreted as the two regions occurring in the study of unstable wetting fronts in sands (Glass et al., 1989), matrix and cracks in clay soils (Bouma, 1980) or matrix and fissures in rocks (Barenblatt et al., 1960).

The *particles* are assumed to be pointlike with no interaction with each other. Physically, they can be interpreted as Lagrangean elements of solute moving in a steady water phase. The N particles move in the 3-dimensional space $\{z,\xi,t\}$ spanned by the coordinates depth z, structure ξ, and time t. In each structure the *movement* of the particles is a Markov process described by a mean velocity v_i and a diffusion coefficient D_i (Gardiner, 1983), where the index i (i=1,2) refers to the structure. A particle is in the microstructure, i=1, for $0 \leq \xi \leq \vartheta_1$ and in the macrostructure, i=2, for $\vartheta_1 < \xi \leq 1$. It diffuses in structure-space [0,1] with diffusion coefficient γ and is reflected at the boundaries. The diffusion coefficient determines the coupling between micro- and macrostructure.

3.2 Implementation

To simulate the movement of the particles through the random medium, time is discretized in steps of Δt and the velocity of the particles is assumed constant during Δt and uncorrelated for different time steps.

The position $\mathbf{x}_j(t_k)$ of particle j at the discrete times t_k is given by the vector

$$\mathbf{x}_j(t_k) := \{\, z_j(t_k)\ ,\ \ \xi_j(t_k) \,\}\ . \tag{1}$$

At time t_0, particle j is at position

$$\mathbf{x}_j(t_0) := \{\, 0\ ,\ \nu \,\}\ , \tag{2}$$

where ν is a uniform random number in the interval [0,1]. Its propagation in physical space (convection and diffusion) and in structure-space (diffusion only) is simulated by a random walk (Gardiner, 1983)

$$\mathbf{x}_j(t_{k+1}) := \left\{\, z_j(t_k) + \ v_i\,\Delta t + \omega\sqrt{6D_i\,\Delta t}\ ,\ \ \xi_j(t_k) + \omega\sqrt{6\gamma\Delta t} \,\right\}\ , \tag{3}$$

where v_i and D_i are mean velocity and diffusion coefficient in structure i, ω is a uniformly distributed random number in the interval [-1,1] and γ is the diffusion coefficient in structure-space. (The factor 6 in front of the diffusion coefficients allows their interpretation as conventional diffusion coefficients on a macroscopic time scale as will be show below.) The position of particle j at time t_k in structure-space determines which set of transport parameters will be applied during the interval $[t_k, t_{k+1}]$:

$$i = \begin{cases} 1 & ;\ 0 \leq \xi_j(t_k) \leq \vartheta_1 \\ 2 & ;\ \vartheta_1 < \xi_j(t_k) \leq 1 \end{cases}\ , \tag{4}$$

where ϑ_1 is the fraction of the transport volume belonging to the microstructure.

3.3 Limiting Cases

Two limiting cases of the diffusion coefficient in structure-space, $\gamma \to 0$ and $\gamma \to \infty$, allow a simple description of the model's behaviour. The procedure for developing the description of the limiting cases consists in deriving a transition probability in physical space from z' to z between time t and t+Δt, $p(z,t+\Delta t \mid z',t)$. Since the model's propagator (equation (3)) is independent of absolute position in space and time, the transition probability is a function of relative positions, only. It is therefore sufficient to calculate $p(z,\Delta t \mid 0,0)$. By invoking the central limit theorem (Papoulis, 1984), expectation value and variance of the Gaussian transition probability $p(z,\tau \mid 0,0)$ on the macroscopic time scale $\tau >> \Delta t$ are calculated from the elementary probability $p(z,\Delta t \mid 0,0)$. This macroscopic transition probability leads finally to a Fokker-Planck-equation for the probability of the particle positions (Gardiner, 1983).

Limit $\gamma \rightarrow 0$: In this limit the double porous medium is the sum of two non-interacting homogeneous media. The probability for a transition from (0,0) to $(z,\Delta t)$ in structure i is

$$p_i(z,\Delta t \mid 0,0) = \begin{cases} \left(2\sqrt{6 D_i \Delta t}\right)^{-1} & ; \left|z - v_i \Delta t\right| \leq \sqrt{6 D_i \Delta t} \\ 0 & ; \left|z - v_i \Delta t\right| > \sqrt{6 D_i \Delta t} \end{cases} . \tag{5}$$

By the central limit theorem the transition probability on a macroscopic time scale $\tau >> \Delta t$ can be approximated by a Gaussian with expectation $\mu_i = v_i \tau$ and variance $\sigma_i^2 = 2 D_i \tau$,

$$p_i(z,\tau \mid 0,0) = \frac{1}{\sqrt{4\pi D_i \tau}} \exp\left[\frac{(z - v_i \tau)^2}{4 D_i \tau}\right]. \tag{6}$$

It can be shown (Gardiner, 1983; Chap. 3) that this transition probability leads to the Fokker-Planck-equation

$$\frac{\partial}{\partial t} f_i(z,t) = -\frac{\partial}{\partial z}\left(v_i f_i(z,t) - D_i \frac{\partial}{\partial z} f_i(z,t)\right) \tag{7}$$

for the pdf $f_i(z,t)$ of the particle positions in structure i. The solutions of equation (7) are also good descriptions of the behaviour of weakly interacting structures as long as

$$\gamma << \frac{\left(\min(\vartheta_1, 1-\vartheta_1)\right)^2}{6\,T} , \tag{8}$$

where T is the total simulation time. Inequality (8) assures that only a very small number of particles in the immediate neighbourhood of the boundary between the structures have a chance to cross it.

Limit $\gamma \rightarrow \infty$: In this limit the double porous medium appears to be homogeneous because the position of a particle in structure-space is a red noise process. (The spectrum of this process is limited by the finite time step Δt of the simulation during which the position of a particle in structure-space does not change.) The transition probability from (0,0) to $(z,\Delta t)$ is the sum of the transition probabilities in the two structures weighted by their volume fraction

$$p(z,\Delta t \mid 0,0) = \vartheta_1\, p_1(z,\Delta t \mid 0,0) + \left(1-\vartheta_1\right) p_2(z,\Delta t \mid 0,0) , \tag{9}$$

where p_1 and p_2 are given by equation (5). A lengthy calculation (Jury and Roth, 1990) shows that on a macroscopic time scale $\tau >> \Delta t$ the transition probability is approximated by the Gaussian

$$p(z,\tau \mid 0,0) = \frac{1}{\sqrt{4\pi \bar{D} \tau}} \exp\left[-\frac{(z - \bar{v}\tau)^2}{4 \bar{D} \tau}\right] , \tag{10}$$

with the effective parameters

$$\bar{v} = \vartheta_1 v_1 + (1-\vartheta_1) v_2 \tag{11a}$$

$$\bar{D} = \left[\vartheta_1 D_1 + (1-\vartheta_1) D_2\right] + \Delta t\left(\frac{1}{2} - \vartheta_1(1-\vartheta_1)\right)(v_1 - v_2)^2 . \tag{11b}$$

These effective parameters of the entire medium are essentially the mean of the corresponding parameters in the single structures weighted by their volume fraction. In addition, the effective diffusion coefficient contains a term which is proportional to the time step Δt of the simulation. This term accounts for the small-scale temporal correlation of the positions in structure-space introduced by the discretization of time. The discrete simulation thus gives a faithful picture of the corresponding continuous process if Δt is chosen so that the second term in equation (11b) is negligible compared to the first one. The transition probability (10) leads to the Fokker-Planck-equation

$$\frac{\partial}{\partial t} f(z,t) = -\frac{\partial}{\partial z}\left(\bar{v}\, f(z,t) - \bar{D}\frac{\partial}{\partial z} f(z,t)\right) \tag{12}$$

for the pdf f(z,t) of the position of the particles in the entire medium. The solution of equation (12) is also a good approximative description of strongly interacting structures with finite diffusion coefficient γ as long as

$$\gamma >> \frac{\left(\max(\vartheta_1, 1-\vartheta_1)\right)^2}{6\,\Delta t} . \tag{13}$$

Inequality (13) assures that most of the particles travel large distances in structure-space during the time step Δt of simulation. They are thus crossing each structure many times during Δt, thereby destroying any correlations between successive time steps.

3.4 Simulations

Finite Interaction between Structures: For many boundary conditions, equations (7) and (12) can be solved analytically. No simple solutions for f(z,t) are available for the general case of a finite interaction between the structures when neither of the two inequalities (8) and (13) is applicable.

To investigate the model in this intermediary region, the particle distribution at time T=1 as a function of depth and diffusion coefficient γ was estimated with a random walk simulation in a medium with $\vartheta_1=0.8$ (fig. 2). The simulation reproduces the analytical results for the two limiting cases $\gamma \rightarrow 0$ and $\gamma \rightarrow \infty$. For $\gamma << 6.7 \cdot 10^{-3}$, the condition required by equation (8), the structures are virtually decoupled and the single input pulse splits into two independent pulses. For $\gamma >> 1.1 \cdot 10^2$, equation (13) with $\Delta t=0.001$, the two structures interact strongly and the medium appears to be homogeneous.

The penetration depth in the microstructure shows a maximum for $\gamma^* \approx 10^{-1}$. This maximum is caused by particles that were initially in the macrostructure and subsequently diffused into the microstructure. The number of particles diffusing from the macro- into the microstructure is very small for $\gamma << \gamma^*$ and does not change the penetration depth in the microstructure appreciably.

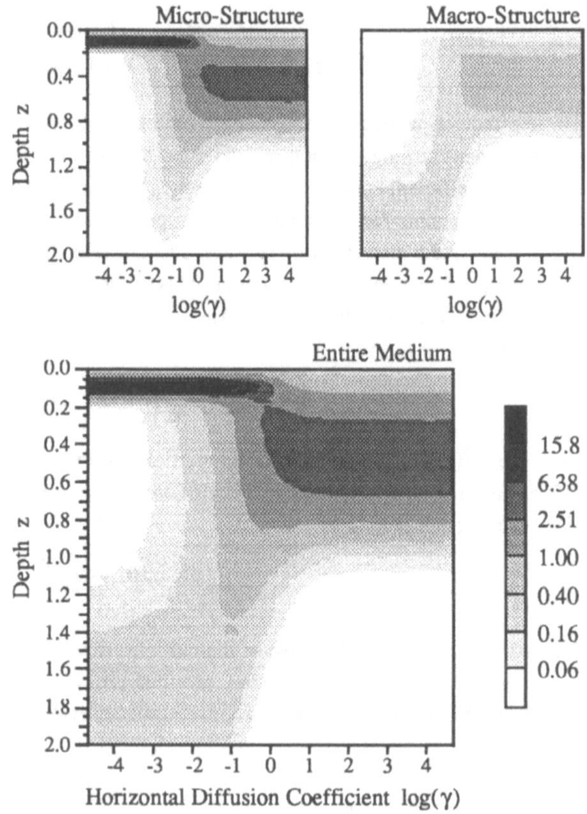

Figure 2: Particle density at time T=1 in micro- and macrostructure and entire medium as a function of depth z and diffusion coefficient γ from the simulation of 2^{16} particles in a dimensionless system. The particle density is normalized with the density of a uniform distribution. Parameters used are $\vartheta_1=0.8$, $v_1=0.1$, $v_2=2.0$, $D_1=10^{-4}$, $D_2=10^{-1}$ *and the time step is* $\Delta t=0.001$. *The effective parameters for* $\gamma\to\infty$ *are* $\bar{v}=0.48$ and $\bar{D}=2\cdot10^{-2}$. *(Note logarithmic scale of gray levels.)*

Conversely, for $\gamma >> \gamma^*$, the particles do not stay long enough in the macrostructure to be transported over a longer distance until they diffuse back into the microstructure. This again leads to a smaller penetration depth in the microstructure then for γ^*.

It is clear from comparing the simulated pdfs in the micro- and the macrostructure, that a measurement in the microstructure is representative for the entire medium only if the interaction between the two structures exceeds a certain value. The particle density in the microstructure is of particular importance because—recalling the experiment described above—it reflects the measurements of suction cups.

Influence of Impermeable Layer: In a next step an impermeable layer is introduced. It is simulated as a sudden increase of the fraction of the transport volume belonging to the microstructure at the depth z=0.8 from $\vartheta_1=0.8$ to $\vartheta_1=0.96$. This sudden change of ϑ_1, as indeed any gradient in ϑ_1, leads to an implicit interaction between the two structures even without an explicit interac-

Figure 3: Geometry for the simulation of an impermeable layer at z_1. The increase of the volume fraction of the microstructure, ϑ_1, leads to an implicit interaction between the two structures at the depth z_1, because part of the particles are forced into the microstructure by the vertical movement in physical space.

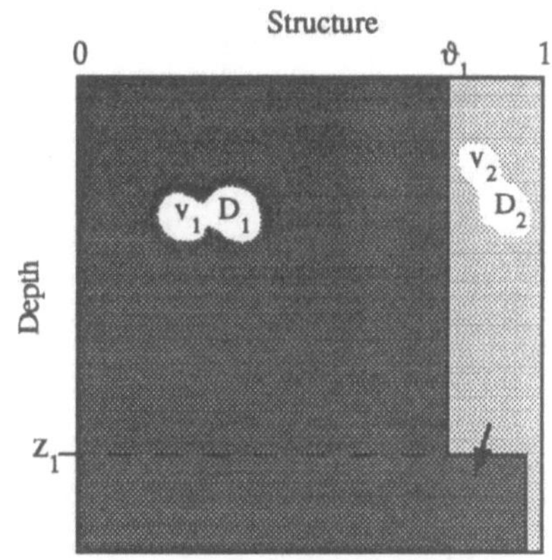

tion, γ=0, as some of the particles cross the boundary between the structures by just a vertical movement (fig. 3).

To illustrate the influence of an impermeable layer, the particle density is simulated for a specific value of the diffusion coefficient in structure-space (fig. 4). The impermeable layer intercepts part of the particles moving in the macrostructure and forces them into the microstructure. This results in the occurrence of a second pulse in the microstructure, seemingly originating at the the upper boundary of the impermeable layer. A comparison of the simulated particle density in the microstructure (the density representing measurements with suction cups) with the experimental mean concentration profile (fig. 1) shows good qualitative agreement.

4. Conclusion and Outlook

It is concluded from the results of the tracer experiment and the subsequent qualitative simulation that transport of conservative solutes under field conditions is a complicated process that may involve splitting of originally single pulses. Measurements with suction cups must be interpreted with great care as they may not be representative for the entire medium. Impermeable layers in the soil profile, or indeed any other hydraulic discontinuities increase the interaction between different structures in the soil and may thereby be very valuable in detecting transport through structures not directly accessible by measurements.

While not quantitative yet, the proposed random walk model offers a simple explanation for the field tracer experiment. It is also a valuable tool for studying transport processes in natural soils theoretically. Future work on this model is directed towards establishing a differential equation that describes the model concepts and towards solving it for some limiting cases. A comparison between one of these limiting cases, v_1=0=D_1, and the mobile-immobile model (van Genuchten and Wierenga, 1976) will clarify the underlying concepts of the two approaches.

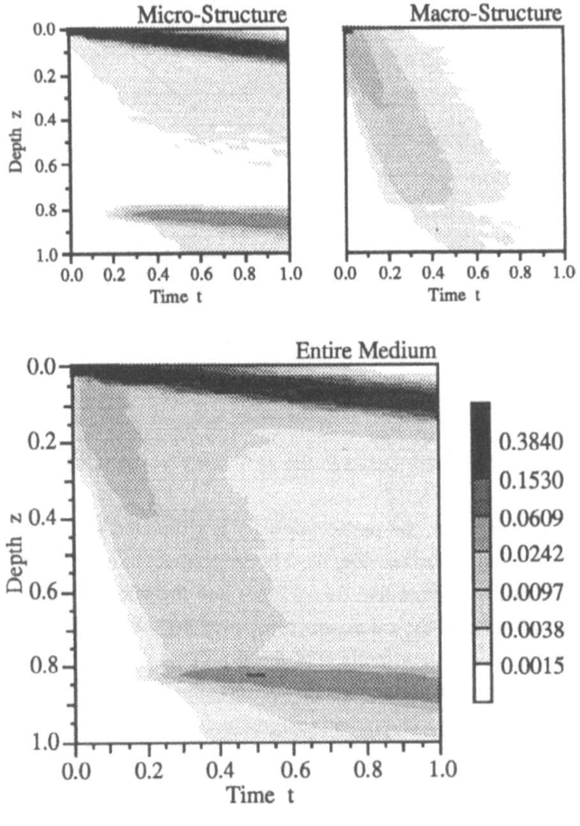

Figure 4: Particle density in a medium with an impermeable layer as a function of depth and time. The impermeable layer is simulated by a jump of ϑ_1 from 0.8 to 0.96 at z=0.8. The diffusion coefficient in structure-space is γ=0.01.

The proposed random walk can easily be extended ion several directions: (i) a multidimensional structure-space allows the simulation of transport in a multiple porous structure, (ii) a continuous dependence of the transport parameters would model spatial variability (Jury and Roth, 1990), (iii) different diffusion coefficients γ_i in the structures i could simulate adsorption, (iv) interacting particles would allow chemical reactions, and (v) non-point particles could be used to simulate non-steady water flow.

Encouraging and clarifying discussions with W.A. Jury about the random walk model are gratefully acknowledged.

References

Aifantis, E.C., 1979: A new interpretation of diffusion in high-diffusivity paths—A continuum approach. Acta metall., **27**, 683-691.

Barenblatt, G.I., I.P. Zheltov, and I.N. Kochina, 1960: Basic concepts in the theory of seepage of homogeneous liquids in fissured rocks [strata]. J. Appl. Math. Mech., **24**, 1286-1303.

Beese, F. and P.J. Wierenga, 1980: Solute transport through soil computed with a transient and a constant flux model. Soil Sci., **129**, 245-253.

Bouma, J., 1980: Field measurement of soil hydraulic properties characterizing water movement through swelling clay soils. J. Hydrol., **45**, 149-158.

Gardiner, C.W., 1983: Handbook of Stochastic Methods for Physics, Chemistry and the Natural Sciences. Springer-Verlag, Berlin.

Glass, R.J., J.-Y. Parlange, and T.S. Steenhuis, 1989: Wetting front instability. I. Theoretical discussion and dimensional analysis. Water Resour. Res., **25**, 1187-1194.

Jury, W.A., H. Elabd, and M. Resketo, 1986: Field study of Napropamide movement through unsaturated soil. Water Resour. Res., **22**, 749-755.

Jury, W.A. and K. Roth, 1990: Transfer functions and solute movement through soil. Theory and application. Birkhäuser Verlag, Basel.

Papoulis, A., 1984: Probability, Random Variables, and Stochastic Processes. Second edition, McGraw-Hill, Auckland.

Roth, K., 1989: Stofftransport im wasserungesättigten Untergrund natürlicher, heterogener Böden unter Feldbedingungen. Diss. ETH Zürich Nr. 8907.

Roth, K., H. Flühler, and Ch. Gysi, 1988: Temporal change of the spatial distribution of chloride displaced through a heterogeneous soil. In *Proceedings of the International Conference on the Validation of Flow and Transport Models for the Unsaturated Zone*, edited by P. Wierenga and D. Bachelet.

Sposito, G., V.K. Gupta, and R.N. Bhattacharya, 1979: Foundation theories of solute transport in porous media: a critical review. Adv. Water Res., **2**, 59-68.

van Genuchten, M.Th. and P.J. Wierenga, 1976: Mass transfer studies in sorbing porous media. I. Analytical solutions. S. Sci. Soc. Am. J., **40**, 473-480.

Walker, G.R., 1987: Solution to a class of coupled linear partial differential equations. IMA J. appl. Math., **38**, 35-48.

Kurt Roth, Hannes Flühler, and Werner Attinger, Soil Physics, ETH Zentrum NO, CH-8092 Zürich

Field-Scale Water and
Solute Flux in Soils
Monte Verità

MASS FLUX OF SORPTIVE SOLUTE IN HETEROGENEOUS SOILS

G. Destouni and V. Cvetkovic

Solute transport in heterogeneous soils is described using the expected mass flux across a surface orthogonal to the mean flow. The expected mass flux is defined for nonreactive and sorptive solute in the general case of a three-dimensional, heterogeneous porous media. In the case of field-scale solute transport through the unsaturated zone it is shown that the influence of joint variability in the hydraulic and sorption parameters on the expected mass arrival at a given depth from the surface may be important. Specifically, a nonequilibrium effect that significantly influences the form of the expected mass arrival may appear when the sorption parameters are spatially variable.

1. Introduction

Transport of pollutants in the subsurface is a result of complex interaction between hydraulic and chemical phenomena. Spatial variability in porous medium properties, such as hydraulic conductivity, has been found to play an important role in the spreading of pollutants both in the saturated and unsaturated porous media (e.g. Bresler and Dagan, 1981; Dagan, 1982; Gelhar and Axness, 1983). Furthermore, the chemical reactions between pollutants and the soil matrix may greatly influence the solute movement (Rubin, 1983). Although these reactions are in general complex, the equilibrium and nonequilibrium sorption-desorption models have provided suitable approximations for surface chemical reactions in groundwater (Travis and Etnier, 1981; Nielsen et al., 1986). In addition, diffusive mass transfer resistances associated with immobile regions, such as intra-aggregate porosity, sorbent organic matter, or low hydraulic conductivity zones, are often significant and are manifested on the field scale as sorption-desorption reactions (Brusseau and Rao, 1989).

Investigations of solute transport in heterogeneous porous media have mainly focused on nonreactive pollutants (Bresler and Dagan, 1981; Dagan, 1982; Gelhar and Axness, 1983; Neuman et al., 1987; Shapiro and Cvetkovic, 1988). Two main approaches have been used in these investigations: the Eulerian approach (Gelhard and Axness, 1983; Neuman et al., 1987), and the Lagrangian approach (Dagan, 1982; 1984; Shapiro and Cvetkovic, 1988). In the Lagrangian approach, one focuses the analysis either on the spatial moments of a pollutant, wherefrom the concentration is defined (Dagan, 1982; 1984), or, on the travel time moments, wherefrom the mass flux is defined (Shapiro and Cvetkovic, 1988; Cvetkovic and Shapiro, 1990). In many instances, the travel time moments and the mass flux of a pollutant are quantities of direct practical interest; advantages of using the mass flux when modelling subsurface contamination have been noted by Dagan and Nguyen (1989).

In this paper, a probabilistic model for the expected mass flux of solute advected in heterogeneous porous media, is discussed. The results obtained by Shapiro and Cvetkovic (1988), and Cvetkovic and Shapiro (1990), are first briefly summarized. Followingly, the analysis focuses on field-scale transport in the unsaturated zone. In particular, the influence of spatial variability in both the hydraulic and sorption-desorption parameters on the expected mass flux through heterogeneous fields, is investigated.

2. Nonreactive solute

We consider a solute particle that is chemically and dynamically inert, and is released into a heterogeneous porous medium at t=0, and x=0; the discussion on the nature of a solute particle is given, for instance, by Dagan (1982, 1984). In order to focus the analysis on solute advection, we shall neglect local dispersion and molecular diffusion.

If the mean velocity is oriented in the x_1 direction, the arrival time of the solute particle at a plane x_1 can be defined in terms of the fluid velocity (Shapiro and Cvetkovic, 1988)

$$T(x_1) = \int_0^{x_1} \frac{d\xi}{w_1(\xi)} \tag{1}$$

where

$$w_1(x_1) = v_1(x_1, X_2(T(x_1)), X_3(T(x_1))) \tag{2}$$

with $X_i = X_i(t)$ (i=1,2,3) being the components of the particle pathline vector, and v_i is the fluid velocity vector. The velocity vector $w_i(x_1)$ is similar to the Lagrangian velocity in that it is related to the particle pathline; however, the independent variable associated with the Lagrangian velocity is time, while the independent variable associated with w_i is the spatial position along the mean flow direction.

The mass flux of the solute particle across a plane perpendicular to the mean fluid motion is $s(x_1,t) = m\delta(t-T(x_1))$, where $\delta()$ is the Dirac delta function and m is the mass of the solute particle. Due to the random character of the flow field, the mass flux is a random function. We may evaluate the expected mass flux across the plane x_1 as $\bar{s} \equiv \langle s \rangle = m p_T(t;x_1)$, where p_T is a probability density function for the arrival time of the particle at x_1. The first two moments of the arrival time are related to the statistical properties of the hydraulic conductivity for three-dimensional, heterogeneous porous media in Shapiro and Cvetkovic (1988).

3. Sorptive solute

Next, we consider a solute particle that is introduced into the porous medium at t=0 and **x**=**0** and is assumed to undergo a nonequilibrium sorption-desorption reaction governed by first-order linear kinetics. The properties of the fluid velocity field are assumed to be equivalent to those discussed for the nonreactive solute particle. If local dispersion and molecular diffusion are neglected, a distribution of the mobile and immobile solute mass exists only along the streamline associated with the particle. The spatial and temporal distribution of the mobile and immobile concentrations of the solute are governed by

$$\frac{\partial C}{\partial t} + \nabla \cdot \mathbf{S} = -\frac{\partial C^*}{\partial t} \tag{3a}$$

$$\frac{\partial C^*}{\partial t} = \kappa_1 C - \kappa_2 C^* \tag{3b}$$

where $C(\mathbf{x},t)$ and $C^*(\mathbf{x},t)$ are the mobile and immobile concentrations in unit volume of the bulk flow system, respectively, $\mathbf{S}(\mathbf{x},t)$ is the mass flux vector of the mobile solute, and $\kappa_1(\mathbf{x})$ and $\kappa_2(\mathbf{x})$ are the sorption and desorption rate coefficients, respectively.

Sorption-desorption models that have been considered in the literature, such as surface adsorption models and mobile-immobile fluid models, are mathematically analogous to (3) and the interpretation of the coefficients κ_1 and κ_2 will depend on the particular model that is hypothesized (Lassey, 1988; Brusseau and Rao, 1989). If $\kappa_2=0$, one obtains the case of solute degradation. If κ_1 and κ_2 become large while κ_1/κ_2 remains finite, the equilibrium linear sorption model is obtained.

Integrating (3) over the plane x_1 and using the fact that the sorption process takes place along a streamline, a system of equations analogous to (3) can be derived for the mass flux as (Cvetkovic and Shapiro, 1990)

$$\frac{\partial s}{\partial t} + w_1(x_1)\frac{\partial s}{\partial x_1} = -\frac{\partial s^*}{\partial t} \tag{4a}$$

$$\frac{\partial s^*}{\partial t} = k_1(x_1)\, s - k_2(x_1)\, s^* \tag{4b}$$

where s is the mass flux in the direction x_1 integrated across the plane at x_1, and $s^* \equiv w_1 c^*$, with c^* being the immobile concentration integrated over x_1. The rate coefficients k_j (j=1,2) are defined as

$$k_j(x_1) = \kappa_j\left(x_1, X_2\left(T(x_1)\right), X_3\left(T(x_1)\right)\right) \tag{5}$$

Equations (4) provide a one-dimensional Eulerian description of the mass flux written in terms of the three-dimensional advection of a sorptive solute particle in a heterogeneous porous medium. The one-dimensional description in (4) retains the three-dimensional character of the transport process by using coefficients w_1, k_1 and k_2 that are tied to a particular streamline, and thus are Lagrangian in nature; the distance along the mean flow direction, x_1, is the parameter associated with these coefficients.

3.1 Expected Mass Flux

The spatial variability in the properties of the porous medium implies that the fluid velocity, $\mathbf{v}(\mathbf{x})$, and the sorption coefficients, $\kappa_j(\mathbf{x})$ (j=1,2), are random fields. Thus, (4) cannot be solved in a deterministic manner. Instead, the expected mass flux is to be defined as a means of quantifying the transport process. A general expression for the expected mass flux can be written in the form (Cvetkovic and Shapiro, 1990)

$$\bar{s} \equiv \langle s \rangle = m \left\langle L^{-1}\left[\exp(-qT)\exp\left(-\int_0^{x_1} \frac{k_1 q\, d\xi}{(q+k_2)\, w_1}\right)\right]\right\rangle \tag{6}$$

where L^{-1} denotes the inverse Laplace transform, q is the Laplace transform variable, and T is defined in (1).

For the special case when κ_1 and κ_2 are constant, the general expression (6) yields

$$\bar{s} = <s(t,T)> = \int_0^{\infty} s(t,T)\, p_T(T)\, dT \tag{7}$$

where s is obtained as the solution of (4) when k_1 and k_2 are constant, and p_T is the arrival time pdf for nonreactive solute. The solution of (4) has been given by Lassey (1988) for an instantaneous injection in the flux and for constant advection and sorption coefficients. From the flux-averaged concentration, the mass flux is:

$$s(t,T) = m \exp(-k_1 t)\, \delta(t-T) + m\, k_1 k_2 T \exp(-k_1 T - k_2 t + k_2 T)\, \hat{I}_1[k_1 k_2 T (t-T)] \tag{8}$$

where $\hat{I}_1(Z) \equiv I_1(2\sqrt{Z})/\sqrt{Z}$, with I_1 being a modified Bessel function of the first kind of order one. The solution (8) is identical for constant and variable advection, provided that k_1 and k_2 are constant (Cvetkovic and Shapiro, 1990).

4. Heterogeneous soils

The spatial heterogeneity in natural fields may have a significant effect on field-scale solute transport in the unsaturated zone (e.g. Bresler and Dagan, 1981; Destouni and Cvetkovic, 1989). Furthermore, preferential flow paths and generally high levels of sorbent organic-matter content are typical conditions in the unsaturated zone that may lead to nonequilibrium sorption-desorption of solute at the field scale (Brusseau and Rao, 1989). Thus, when modelling field-scale mass flux of solute in the unsaturated zone there is a need to couple sorption-desorption with the variability in different hydraulic and sorption parameters.

We consider an areal source of contamination at the soil surface of a field where the horizontal heterogeneity is dominant. Water flow takes place due to a constant rate of recharge R, at the soil moisture content θ, and at the pore water velocity v. We assume vertical gravitational steady flow and vertically homogeneous soil. The pore water velocity v is then related to the soil hydraulic properties and the recharge boundary conditions in the following way (Bresler and Dagan, 1981):

$$v = \frac{K(\theta)}{\theta} = \frac{R}{\theta} \quad \text{for } R < K_s \tag{9a}$$

$$v = \frac{K_s}{\theta_s} \quad \text{for } R \geq K_s \tag{9b}$$

where K_s and θ_s are the hydraulic conductivity and the water content at saturation, respectively. The relationship $K(\theta)$ adopted here is

$$K(\theta) = K_s \left(\frac{\theta - \theta_{ir}}{\theta_s - \theta_{ir}} \right)^{1/\beta} \tag{10}$$

where θ_{ir} is the irreducible water content and β is a soil coefficient related to the pore size distribution; in the following we assume $\theta_{ir}=0$ and $\beta=0.14$.

In view of the above flow assumptions, the linear nonequilibrium transport model given in (4) may be applied in each profile with the mean pore water velocity v, independent of the depth z, i.e.

$$\frac{\partial s}{\partial t} + v\frac{\partial s}{\partial z} = -\frac{\partial s^*}{\partial t} \tag{11a}$$

$$\frac{\partial s^*}{\partial t} = k_1 s - k_2 s^* \tag{11b}$$

where s is the mass flux in the profile, $s^*=vc^*$, and k_1 and k_2 are the sorption parameters; in this case $k_j \equiv \kappa_j$ (j=1,2). The solution of (11) for an instantaneous injection in the flux, in a semi-infinite domain, is given by (8) with T=z/v.

We regard the saturated hydraulic conductivity K_s, as a lognormal random variable. Furthermore, we shall assume perfect correlation between the sorption parameters k_1, k_2, and K_s, i.e.

$$K_s = K_s^G \exp(Y) \tag{12}$$

$$k_1 = k_1^G \exp(-Y) \tag{13}$$

$$k_2 = k_2^G \exp(-Y) \tag{14}$$

where K_s^G, k_1^G and k_2^G are the geometric means of K, k_1 and k_2, respectively, and Y is a normally distributed random variable with zero mean and variance σ^2. Thus, K_s and k_1, k_2 are assumed negatively correlated.

In the unsaturated zone, at the field scale, physical nonequilibrium is likely to be the predominant nonequilibrium causing mechanism (Brusseau and Rao, 1989). In that case the negative correlation assumption implies that for an increasing hydraulic conductivity in the mobile zone the solute transfer rate between the mobile and the immobile regions decreases. This assumption is consistent with the fact that, on the one side, a decrease in the surface-to-volume ratio leads to a reduction in solute transfer rate (Rasmuson, 1985), while on the other, a lower surface-to-volume ratio generally implies a higher hydraulic conductivity. For instance, the mobile region may represent the macropores in a given field and in such a case a lower tortuosity of a macropore results both in a lower surface-to-volume ratio and a higher hydraulic conductivity.

Since all parameters depend on the random variable Y, the expression for the expected mass flux can be written as

$$\bar{s} = \int_{-\infty}^{\infty} s\,(z,t;Y) p_Y(Y)\, dY \tag{15}$$

Based on the ergodic hypothesis (Bresler and Dagan, 1981), we assume the expected field-scale mass flux to be approximately equal to the ensemble average given by (15).

Figures 1-2 illustrate the expected field-scale cumulative mass arrival at a given depth below the soil surface for a nonreactive and a sorptive solute, respectively, when only K_s varies. The cumulative mass arrival is determined as the time integral of the mass flux. Different magnitudes of variability are considered. Comparison between Figures 1 and 2 indicates that nonequilibrium sortion-desorption decreases the effect of spatial variability in the hydraulic conductivity on field-scale solute transport. Figure 3 shows cumulative mass arrival curves when only K_s varies compared with the corresponding curves when K_s, k_1 and k_2 vary simultaneously. For increasing mean values of the sorption-desorption rate coefficients, the effect of spatial variability in the sorption parameters on the field-scale mass arrival increases. Figure 4 illustrates that this effect may in some cases be significant for the shape of the breakthrough curve. A nonequilibrium effect of asymmetry in conjunction with an early breakthrough is apparent when the sorption parameters vary spatially. This effect does not appear when k_1 and k_2 are constant, because their magnitudes are sufficiently large for the local equilibrium assumption to be valid. The sharp bends that can be seen in all the curves, Figures 1-4, stem from the conditions given in (9a) and (9b).

In this investigation we have restricted the discussion to the case of negative perfect correlation between the hydraulic conductivity and the sorption parameters. Although this is plausible for physical nonequilibrium, other types of correlation could also be considered. For instance, the case of positive perfect correlation, which has been considered in Destouni and Cvetkovic (1990), also results in an altered form of the breakthrough curve in comparison to the case of constant sorption parameters.

Similar effects to those illustrated in Figures 1-4 are found also for a dimensionless formulation of (11), where a specific depth does not have to be assumed, and where equilibrium sorption can be accounted for by a retardation factor (Destouni and Cvetkovic, 1990).

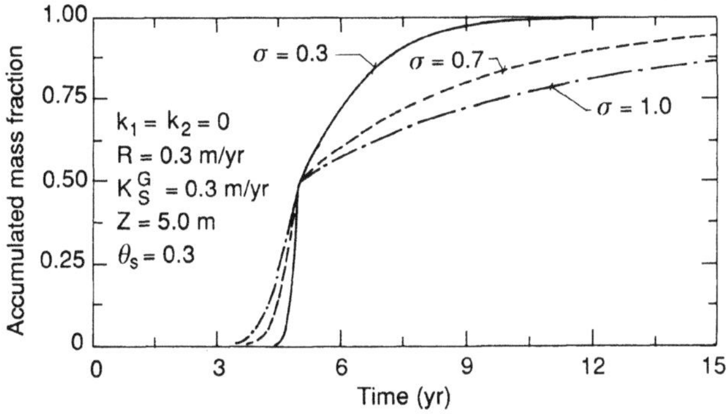

Figure 1: Field-scale cumulative mass arrival for nonreactive solute when only K_S varies.

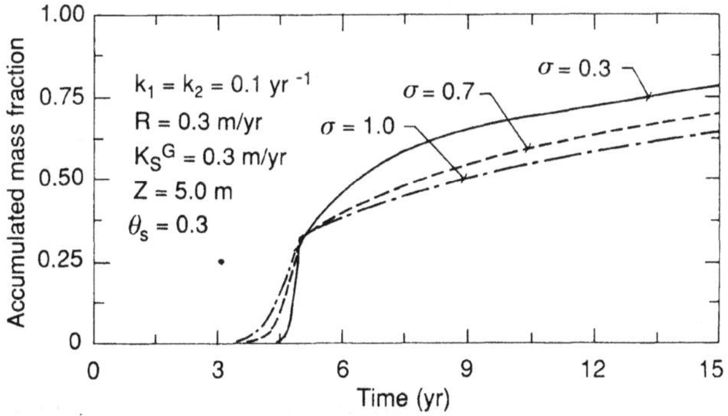

Figure 2: Field-scale cumulative mass arrival for sorptive solute when only K_S varies

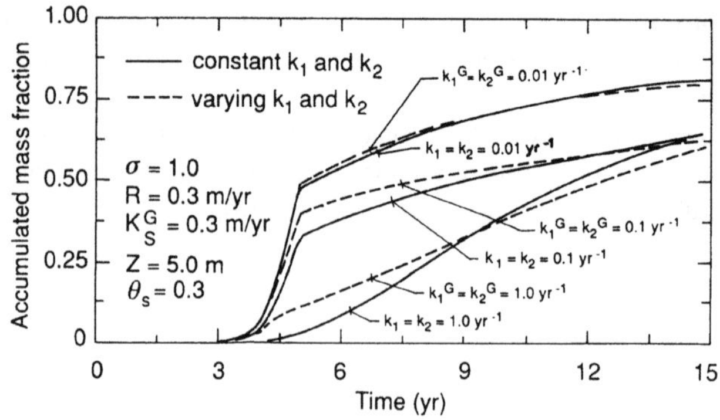

Figure 3: Field-scale cumulative mass arrival for sorptive solute

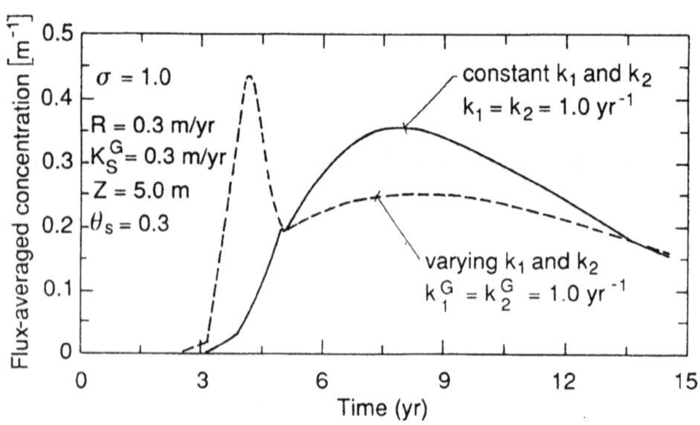

Figure 4: Field-scale flux-averaged concentration for sorptive solute, normalized with the applied mass per unit cross-sectional area

5. Summary and conclusions

Transport of sorptive solute in heterogeneous soils has been described using the arrival time concept. In particular, the expression for the expected mass arrival, or mass flux, across a plane that is orthogonal to the mean flow has been derived for nonreactive solute, and solute undergoing nonequilibrium sorption-desorption with linear kinetics.

Mass flux of sorptive solute through fields where spatial variability in the horizontal plane is dominant, has been analyzed. In particular, the influence of joint variability in the hydraulic conductivity at saturation, and the sorption rate coefficients, on the expected field-scale mass flux, has been investigated. The obtained results indicate that nonequilibrium sorption-desorption decreases the effect of spatial variability in the saturated hydraulic conductivity. Furthermore, a nonequilibrium effect of asymmetry with an early peak may appear in the breakthrough curve if the sorption parameters are spatially variable and negatively correlated with the saturated hydraulic conductivity. Although the present results are primarily qualitative, they indicate that nonequilibrium effects and spatial variability in the sorption parameters may be important when estimating the first arrival of solute into the groundwater through the unsaturated zone.

Acknowledgments

Partial support for this work has been provided by the National Swedish Environmental Protection Board (SNV); the authors in particular acknowledge the assistance of Jan Byman, SNV.

References

Brusseau, M.L., and P.S.C. Rao, 1989: *Sorption nonideality during organic contaminant transport in porous media.* CRC Critical Reviews in Environmental Control, **19**, 33-99.

Bresler, E., and G. Dagan, 1981: *Convective and pore scale dispersive solute transport in unsaturated heterogeneous fields.* Water Resour. Res., **17**, 1683-1693.

Cvetkovic, V.D., and A.M. Shapiro, 1990: *Mass arrival of sorptive solute in heterogeneous porous media.* Water Resour. Res., in press.

Dagan, G., 1982: *Stochastic modeling of groundwater flow by unconditional and conditional probabilities, 2. The solute transport.* Water Resour. Res., **18**, 835-848.

Dagan, G., 1984: *Solute transport in heterogeneous porous formations.* J. Fluid Mech., **145**, 151-177.

Destouni, G., and V. Cvetkovic, 1989: *The effect of heterogeneity on large scale solute transport in the unsaturated zone.* Nordic Hydrology, **20**, 43-52.

Destouni, G., and V. Cvetkovic, 1990: *Field-scale mass arrival of sorptive solute into the groundwater.* Water Resour. Res. (in review).

Gelhar, L.J., and C.L. Axness, 1983: *Three-dimensional stochastic analysis of macrodispersion in aquifers.* Water Resour. Res., **19**, 161-180.

Lassey, K.R., 1988: *Unidimensional solute transport incorporating equilibrium and rate-limited isotherms with first-order loss, 1. Model conceptualizations and analytic solutions.* Water Resour. Res., **24**, 343-350.

Neuman, S., C.L. Winter, and C.M. Newman, 1987: *Stochastic theory of field-scale Fickian dispersion in anisotropic porous media.* Water Resour. Res., **23**, 453-466.

Nielsen, D.R., M.Th. van Genuchten, and J.W. Biggar, 1985: *Water and solute transport processes in the unsaturated zone*. Water Resour. Res., **22**, 89S-108S.

Rasmuson, A., 1985: *The effect of particles of variable size, shape, and properties on the dynamics of fixed beds*. Chem. Eng. Sci., **40**, 621-630.

Rubin, J., 1983: *Transport of reactive solutes in porous media: Relation between mathematical nature of problem formulation and chemical nature of reactions*. Water Resour. Res., **19**, 1231-1252.

Shapiro, A.M., and V.D. Cvetkovic, 1988: *Stochastic analysis of solute arrival time in heterogeneous porous media*. Water Resour. Res., **24**, 1711-1718.

Travis, C.C., and E.L. Etnier, 1981: *A survey of sorption relationships for reactive solutes in soil*. J. Environ. Qual., **10**, 8-17.

G. Destouni, V. Cvetkovic, Hydraulics Engineering, The Royal Institute of Technology,
S-100 44 Stockholm, Sweden.

Field-Scale Water and
Solute Flux in Soils
Monte Verità

EFFECTIVE PROPERTIES FOR MODELING UNSATURATED FLOW IN LARGE-SCALE HETEROGENEOUS POROUS MEDIA

J. L. Zhu, S. Mishra and J. C. Parker

The feasibility of representing unsaturated flow in spatially heterogeneous systems using effective parameters as an "equivalent" homogeneous medium is investigated numerically. Random fields of soil hydraulic properties are indirectly generated from local grain size distribution statistics. Two-dimensional finite element simulations of transient unsaturated flow are conducted for synthetic heterogeneous media. Effective medium properties, derived from a perturbation approximation of Richards' equation, are numerically evaluated and parameterized in terms of the statistics of spatially variable soil properties and mean flow characteristics. Good agreement is noted between the response of heterogeneous and equivalent homogeneous systems.

1. Introduction

Field-scale simulations of unsaturated flow are complicated by the large degree of spatial variability exhibited by natural soil materials. One approach to deal with soil heterogeneity would be to construct a detailed description of the physical domain. This would not only pose formidable demands on the amount of data needed to define the system, but also require enormous computational resources. An alternative modeling approach is to treat the actual heterogeneous medium as an "equivalent" homogeneous system characterized by some effective properties. This requires the development of procedures for defining effective large-scale properties by scaling-up small-scale parameters.

The behavior of effective medium properties for unsaturated flow in heterogeneous media was studied by Yeh et al. (1985) and Mantoglou and Gelhar (1987) using simple parametric models to represent soil hydraulic properties as stationary random processes. They used spectral representation techniques to solve perturbation approximations of the governing stochastic differential equations. An alternative approach would be to perform Monte-Carlo simulations of the unsaturated flow equation numerically for many equiprobable distributed parameter fields of soil hydraulic properties. Although computationally expensive, the Monte-Carlo methodology allows incorporation of boundary effects as well as arbitrary distributions of soil properties. Such analyses pertaining to unsaturated flow in heterogeneous media have been presented by El-Kadi (1987), Hopmans et al. (1988) and Binley et al. (1989). Ababou and Gelhar (1988) presented fine-scale simulations of unsaturated flow in single-realizations of heterogeneous soils.

In this paper our objective is to investigate the relationship between large-scale effective unsaturated flow parameters, local-scale spatially variable soil properties and mean flow characteristics using a hybrid perturbation—Monte-Carlo simulation methodology. Random fields of unsaturated flow parameters will be generated from the statistics of grain size data—assuming that the variability in soil hydraulic properties is principally due to the variability in local particle sizes. Expressions for effective properties will be derived from perturbation approximations of the governing equations of unsaturated flow. Numerical simulations of transient unsaturated flow will be conducted to evaluate suitable parametric forms for the effective properties.

2. Stochastic Flow Equations

The governing equation for transient unsaturated flow in a 2-D vertical domain is given by Richards' equation:

$$C\frac{\partial h}{\partial t}=\frac{\partial}{\partial x}\left(K\frac{\partial h}{\partial x}\right)+\frac{\partial}{\partial z}\left(K\frac{\partial h}{\partial z}+K\right) \tag{1}$$

where h is capillary head, K is unsaturated conductivity, $C = d\theta/dh$ is specific moisture capacity with θ the volumetric water content, t is time, and x and z are horizontal and vertical space coordinates, respectively. At the local scale, K(h) and C(h) are assumed to be described by van Genuchten's (VG) (1980) parametric model such that

$$\theta=\left(\theta_s-\theta_r\right)\left(1+(\alpha h)^n\right)^{-m}+\theta_r \tag{2}$$

$$K=K_s\frac{\left(1-(\alpha h)^{n-1}\left[1+(\alpha h)^n\right]^{-m}\right)^2}{\left[1+(\alpha h)^n\right]^{m/2}} \tag{3}$$

where θ_s the saturated water content, θ_r is a "residual" water content, K_s is the saturated conductivity, α and n are VG model parameters, and the exponent $m = 1-1/n$. The water capacity $C = d\theta/dh$ is obtained by differentiation of Eq. (3).

In a heterogeneous soil, C and K are spatially variable and will be treated as stochastic functions. Anticipating a log-normal distribution of hydraulic conductivity, K(h) can be decomposed as $K(h) = K_g(h)\, e^{f'(h)}$ where K_g is the expected value of $f=\ln K(h)$ (i.e. the geometric mean) and f' represents fluctuation of f about the mean such that $E\{f'\} = 0$. Likewise, the moisture capacity is decomposed as $C(h) = \overline{C}(h)+C'(h)$, where $\overline{C}$ is mean capacity and C' is the fluctuation with $E\{C'\} = 0$ assuming a symmetric distribution. Assuming that stochastic K and C result in h being a stationary stochastic process, we have $h = \bar{h} + h'$ with $E\{h'\} = 0$, $E\{\partial h'/\partial t\} = 0$ and $E\{\partial h'/\partial x_i\} = 0$ where $x_i = x,z$, $\bar{h}$ is the mean head and h' is the fluctuation for the head. Substituting for C, K and h into Eq. (1), taking the expectation of both sides and rearranging, we obtain a stochastic mean flow equation having the same form as the Richards equation, viz

$$C_e \frac{\partial \bar{h}}{\partial t} = \frac{\partial}{\partial x}\left(K_{e_x} \frac{\partial \bar{h}}{\partial x}\right) + \frac{\partial}{\partial z}\left(K_{e_z} \frac{\partial \bar{h}}{\partial z} + K_{e_z}\right) \tag{4}$$

with effective moisture capacity defined by

$$C_e = \bar{C} + \frac{E\{C' \, \partial h'/\partial t\}}{\partial \bar{h}/\partial t} \tag{5}$$

and effective hydraulic conductivity in the x_i-direction defined by

$$K_{e_{xi}} = K_g \left(E\{e^{f'}\} + \frac{E\{e^{f'} \partial h'/\partial x_i\}}{\partial(\bar{h}+z)/\partial x_i} \right) \tag{6}$$

The effective properties defined by Eqs. (4)-(6) exhibit added nonlinearity due to dependence on spatial and temporal derivatives of the mean head, $\bar{h}$. Yeh et al. (1985) and Mantoglou and Gelhar (1987) have used spectral representation techniques to evaluate perturbation derivative terms such as $E\{e^{f'}(\partial h'/\partial x_i)\}$ and $E\{C'(\partial h'/\partial t)\}$ in order to derive analytical solutions for the effective parameters. A numerical method will be used here to solve the flow problem defined by Eq. (1) in conjunction with a Monte-Carlo approach. Details of the two-dimensional Galerkin finite element model with isoparametric elements and forward finite difference approximation in time may be found in Kuo et al. (1989).

3. Generation of Random Fields

The Monte Carlo approach involves solution of Eq. (1) with soil hydraulic properties (i.e. VG model parameters) varying from node to node in the simulation domain. The stochastic parameters of interest (θ_s, θ_r, K_s, α, n) are assumed to be second-order stationary random functions. Random fields of these parameters are generated using an indirect protocol described by Mishra et al. (1989) based on the assumption that variability in soil properties is essentially due to spatial variations in particle size distributions. This procedure involves using a stochastic moving average algorithm to generate spatially variable autocorrelated fields of saturated water content, θ_s, and particle size distribution moments, $\mu_{ln\text{-}d}$ and $\sigma_{ln\text{-}d}$, assuming these variables to be normally distributed with known mean and auto-covariance and with particle diameter, d, log-normally distributed. These randomly generated variables are used to compute the particle size distribution at each node of the simulation domain, from which the stochastic VG model parameters (α, n) and saturated conductivity (K_s) are estimated via physically based pore-structure models as described in detail by Mishra et al. The residual water content, θ_r, was taken to be equal to zero here for convenience. Table 1 shows the statistics of the primary variables ($\mu_{ln\text{-}d}$, $\sigma_{ln\text{-}d}$) and the generated variables (α, n, K_s) for the hypothetical 30x30 node isotropic porous medium used in this study. Predicted cross-correlations were negligible. Computed distributions of α, n and K_s appear log-normal, in agreement with field measurements of soil spatial variability (Hopmans et al., 1988). K(h) also appears to be log-normal for a fixed head and C(h) is symmetrically distributed for a fixed head, confirming the assumptions that lead to the derivation of Eq. (4).

Table 1: Random field generation for hypothetical medium. Units of d(cm), α(1/cm) and K_s(cm/h).

	Parameter	Mean	Std Dev	Corr Length
Input	$\mu_{ln\text{-}d}$	-5.5	0.20	4 blocks
	$\sigma_{ln\text{-}d}$	0.6	0.30	4 blocks
	ln(α)	-4.32	0.521	3.83 blocks
Output	ln(n)	0.58	0.074	3.44 blocks
	ln(K_s)	1.07	0.310	3.84 blocks

4. Evaluation of Effective Properties

The evaluation of effective soil properties given by Eqs. (5) and (6) was carried out as follows. First, $\overline{C}$, K_g, σ_f^2 and $E\{e^{f'}\}$ were computed as functions of h from the local C(h) and K(h) functions by direct averaging over the distributions. Simulations were carried out for 30 realizations of the generated heterogeneous porous media by solving Eq. (1) for the system geometry and boundary conditions shown in Figure 1. For a given time step and a particular element, the following terms were calculated by taking averages over the total number of simulations: (a) mean pressure head, $\overline{h}$, and its derivatives, $\partial\overline{h}/\partial t$ and $\partial\overline{h}/\partial z$, (b) mean water saturation, $\overline{S} = E\{\theta/\theta_s\}$ and (c) the fluctuation terms $E\{e^{f'}(\partial h'/\partial z)\}$. To ensure a large enough sample size, the ergodicity assumption is used, that is, the probability distribution of a single spatial realization is viewed as that of ensemble. Since the mean flow is vertical, averaging was also carried out horizontally (over all 30 nodes) for each simulation. This averaging procedure yields 30 sets of the above variables for each time step with each set representing averaged values at a given depth. Since the prescribed pressure heads along the top and bottom boundaries are uniform, K_{e_x} will not affect the computation of mean flow in the vertical direction – hence $E\{e^{f'}(\partial h'/\partial x)\}$ was not evaluated. Head contours at t=1.1 hr for a single realization are shown in Figure 1.
When the time derivative of the mean head is large, i.e., $\overline{S} = E\{\theta/\theta_s\}$ Eq. (5) suggests that $C_e \cong \overline{C}$. On the other hand, when $|\partial\overline{h}/\partial t| >> 0$, the mean head is not sensitive to water capacity. Furthermore, numerical evaluation of $E\{C'(\partial h'/\partial t)\}$ indicates that it is at least one order of magnitude smaller than the mean capacity, $\overline{C}$, at all times. Thus, the effective capacity, C_e, was taken to be equal to $\overline{C}$ as a good working approximation.

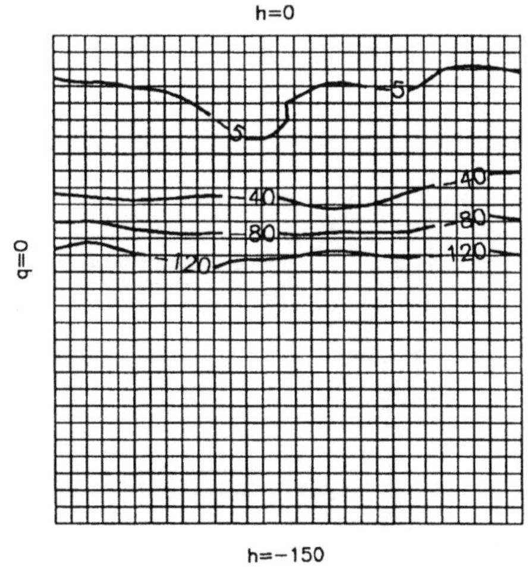

Figure 1: Head contours at t=1.1 hr for a single realization of flow problem in heterogeneous system.

An examination of the spatial and temporal distribution of this mean fluctuation term indicated that it could be approximated by the following empirical expression

$$E\left\{e^{f}\frac{\partial h'}{\partial z}\right\} = \frac{\sigma_f^2}{2}\left\{\exp\left\{-\left(\frac{b-b_m}{b_m}\right)\right\}^2 - 1\right\}(b+1) \qquad (7)$$

where $b = \partial\bar{h}/\partial z$ and b_m=Max|b| where the maximum is taken over the spatial domain at a given time step. Substitution of Eq. (7) into Eq. (6) yields the required expression for effective vertical unsaturated conductivity K_e:

$$K_e = K_g\left(1 + \frac{\sigma_f^2}{2}\exp\left\{-\left(\frac{b-b_m}{b_m}\right)^2\right\}\right) \qquad (8)$$

The spatial distributions of K_e/K_g computed directly from Eq. (6) and its approximation given by Eq. (8) at different time steps are shown in Figure 2. The variable exhibits a peak at the wetting front, where $K_e/K_g \approx 1+\sigma_f^2/2$, while at some distance from the wetting front, $K_e/K_g \approx 1$. This indicates that the effective conductivity is approximated by the arithmetic mean at the wetting front and by the geometric mean at distances sufficiently far from the wetting front. El-Kadi and Brustaert (1985) found that effective conductivity was a function of time. The geometric mean

K_g approximated the effective conductivity at large times, but underestimated the outflow from the aquifer for smaller times. Figure 2 shows that the effective conductivity is a function of both space and time. When soil becomes saturated at large times or is close to steady state, K_g can be taken as the effective conductivity, while at initial transient stages, the arithmetic mean is more relevant where the head gradient changes abruptly.

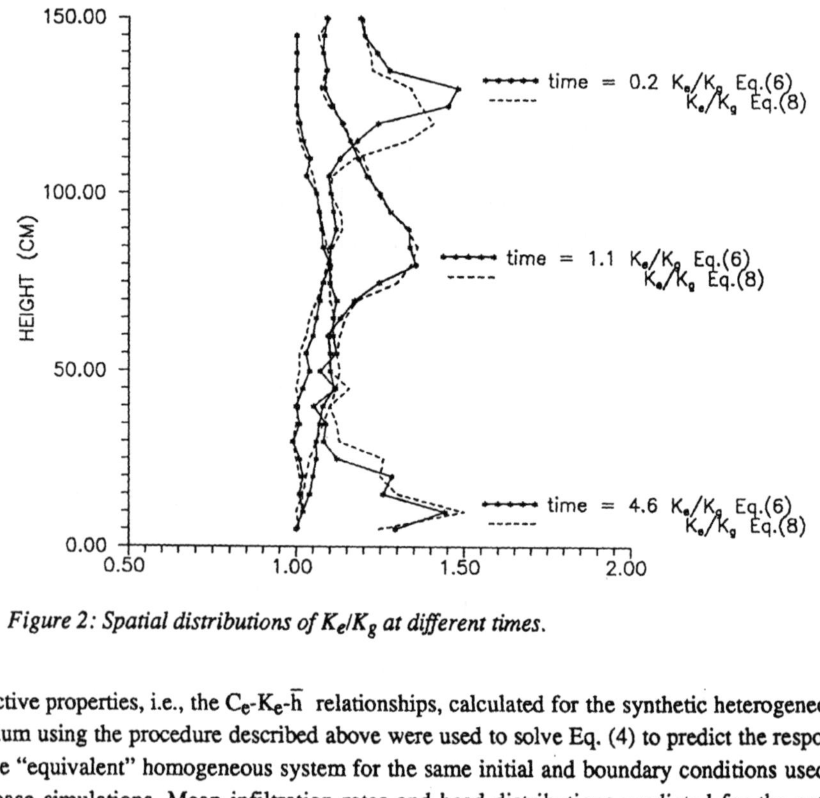

Figure 2: Spatial distributions of K_e/K_g at different times.

Effective properties, i.e., the C_e-K_e-$\bar{h}$ relationships, calculated for the synthetic heterogeneous medium using the procedure described above were used to solve Eq. (4) to predict the response of the "equivalent" homogeneous system for the same initial and boundary conditions used in the base simulations. Mean infiltration rates and head distributions predicted for the actual heterogeneous system using Eq. (1) and for the equivalent homogeneous system using Eq. (4) are compared in Figure 3.

There is good agreement between the response of the heterogeneous system and that of the equivalent homogeneous system both at short times when flow is highly transient as well as at large times as steady state flow is approached. Attempts to approximate the heterogeneous system with hydraulic conductivity represented by the simple geometric mean function $K_g(h)$ was unable to describe the system behaviour accurately (results not shown), indicating that the time-dependent nature of the effective properties must be taken into account to obtain an accurate large scale representation of heterogeneous systems.

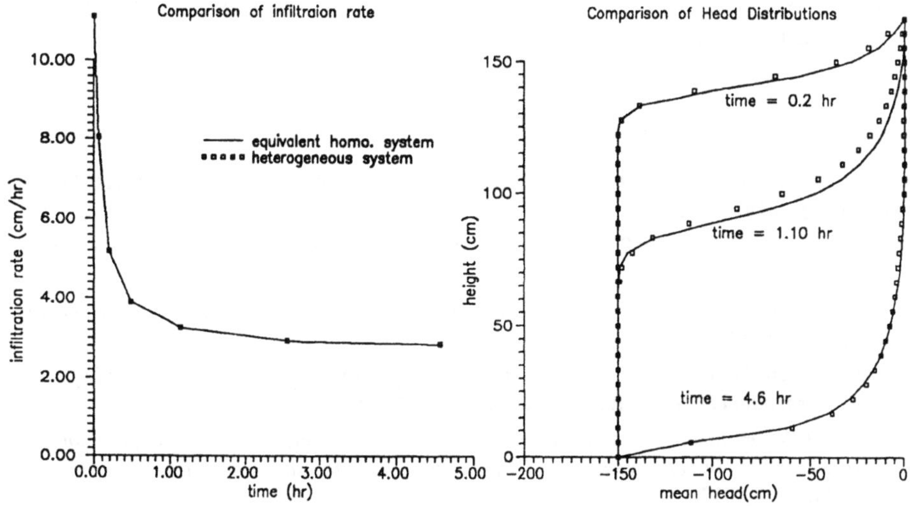

Figure 3: Comparison of head and flux distributions for heterogeneous and equivalent homogeneous systems.

5. Conclusions

A perturbation approach has been used to derive general expressions for effective medium properties. Terms involving expectations of coefficient fluctuations or head fluctuations were evaluated numerically using Monte Carlo simulations and parameterized in terms of soil spatial variability statistics and mean flow attributes. The method was successfully demonstrated for the case of mean 1-D vertical flow by comparison of simulation results for heterogeneous systems and an equivalent homogeneous system defined by effective coefficients. Generalization of the method will require extension to cases of 2-D and 3-D mean flow for systems which may be statistically anisotropic. The practical utility of the method will depend on the feasibility of developing procedures for estimating effective properties from information on local property statistics without general recourse to Monte Carlo simulations. Investigations are currently in progress to develop such a methodology for defining effective properties governing flow in heterogeneous systems.

Acknowledgements

Funds for this work were provided by the USDA-ARS Appalachian Soil and Water Conservation Laboratory under contract 58-3615-7-003.

References

Ababou, R., 1988: *Three-dimensional flow in random porous media*. Ph.D. thesis, M.I.T., Cambridge MA, 2 Vols., pp. 833.

Binley, K. and R. Beven, 1989: *Unsaturated flow in heterogeneous hillslopes*. Water Resour. Res., **25**, 576-587.

El-Kadi, A., 1987: *Variability of infiltration under uncertainty in unsaturated zone parameters*. J. Hydrol., **90**, 61-80.

El-Kadi, A., and W. Brusaert, 1985: *Applicability of effective parameters for unsteady flow in nonuniform aquifers*. Water Resour. Res., **21**(2), 813-198.

Hopmans, J.W., H. Schukking and P.J.J.F. Torfs, 1988: *Two-dimensional steady state unsaturated water flow in heterogeneous soils with autocorrelated soil properties*. Water Resour. Res., **24**, 2005-2017.

Kuo, C.Y., J.L. Zhu and L.A. Dollard, 1989: *A study of infiltration trenches*. Virginia Wate Resources Research Center, Bulletin **163**, 87 pp.

Mantoglou, A. and L.W. Gelhar, 1987: *Stochastic analysis of large-scale transient unsaturated flow systems*. Water Resour. Res., **23**, 37-67.

Mishra, S., J.C. Parker and J.L. Zhu, 1989: *An algorithm for generating spatially autocorrelated unsaturated flow properties*. Comp. Geosc. (in press).

Yeh, T.-C., L.W. Gelhar and A.L. Gutjahr, 1985: *Stochastic analysis of unsaturated flow in heterogeneous soils*. Water Resour. Res., **21**, 447-476.

J. L. Zhu, S. Mishra, and J. C. Parker, Virginia Polytechnic Institute, Blacksburg VA 24061, USA

Field-Scale Water and
Solute Flux in Soils
Monte Verità

TRANSPORT OF REACTIVE SOLUTES IN SPATIALLY VARIABLE UNSATURATED SOILS

S.E.A.T.M. van der Zee

For a solute with adsorption and first order decay an analytical form solution is presented for the field average concentration front when flow properties are random and lognormally distributed. Effects are shown for atrazine displacement with similar characteristic times of decay and of flow. Flow variability affects mainly the short term front. Concentrations are below those of the limiting front for infinite times and breakthrough occurs sooner than expected with the mean residence times. Changes in the retardation factor (R) value affect the short term fronts only. Changes in the decay coefficient μ affect the concentration levels, but not the rate of approach to the limiting front. Variability of R and μ had small effects on the average front for the reference case. For insignificant decay, the effect of variability of R was profound.

1. INTRODUCTION

A large number of contaminants are brought into the soil system by common agricultural practise (fertilization, irrigation, pesticide application) and by unintended pollution. In part these contaminants are immobilized in the soil, but in part they may leach into ground water or be taken up by the vegetation. To assess the hazard of leaching and of adverse effects on ground water quality, sorption, uptake and decay need to be quantified.

The quantification of transport is commonly done by mathematical modeling. With models transport may be predicted for situations that would require much more effort if we had to resort to measurements. Predictions for the field using parameter values found in laboratory research may be poor, because of heterogeneity between e.g. soil samples and field soils. Both the degree of heterogeneity and the scale of heterogeneities differ between these systems. For an accurate extrapolation to the field, the larger scale heterogeneity absent in soil samples and columns should be taken into account. Small scale heterogeneity present in both systems may have too fine a resolution for use in field applications. Therefore appropriate averaging from the microscopic to the macroscopic (field) scale is needed. In this paper transport is modelled for spatially (horizontal) variable soil. It is assumed that the predominant scale of heterogeneity is larger than the cross-sectional scale of soil columns, which are considered homogeneous in the vertical direction and non-interacting with other soil columns. Also we deal with a statistically homogeneous flow domain, i.e., significant trends on the field scale are absent.

2. THEORY

For one dimensional transport in the positive z-direction the transport equation is

$$\frac{\partial}{\partial t}\left[\rho S + \theta c\right] = \frac{\partial}{\partial z}\left[\theta D \frac{\partial c}{\partial z} - \theta v c\right] + \phi(c, S, \ldots) \tag{1}$$

assuming uniformity with depth (z). (Symbols: see Notation). The velocity (v) and θ follow from the solution of the flow equation. Steady state flow is assumed. The hydraulic properties used to solve the flow equation are the retention curve as given by Van Genuchten [1980] and the K(θ)-relationship as given by Mualem [1976]. In the horizontal plane, hydraulic properties may vary significantly in the field. This variability may sometimes be described with the scaling theory of similar media [Warrick et al., [1977]. Denoting a reference (e.g. mean) situation of a scaleable system by the subscript m, all regions are then considered a magnification of this reference, where the ratio (α_i) of characteristic length scales is often lognormally distributed. Scaling theory yields (at a constant value of either θ or water saturation)

$$h_i = \frac{h_m}{\alpha_i}; \qquad K_i = \alpha_i^2 K_m \tag{2}$$

for different realizations (i) a distributed α implies differences in flow in the horizontal plane. To quantify field averaged transport of non-reactive solutes Dagan and Bresler [1979] used scaling theory. Because often the average field response is of more interest than the local transport behaviour, and differs from the response using field average parameter values in the CDE, they evaluated the average concentration with the first moment of the local concentration

$$\bar{c}(z,t) = \int c f(z,t;c) dc \tag{3}$$

where f is the probability density function of c at depth z and time t and assuming steady vertical infiltration with random α and recharge conditions. A closed form solution for (3) was derived by Dagan and Bresler [1979] letting the relative concentration vary between 0 and 1. Pore scale dispersion was neglected. Calculations showed that spatial variability resulted in significant field scale dispersion. Non-Fickian displacement was found which led to the conclusion that a " field equivalent " column with average single valued properties can not (always) be found. In later work the effect of pore scale dispersion was shown to be often less important than spatial variability [Bresler and Dagan, 1981]. Their pioneering approach was used by many others to deal with spatially variable transport.

A related approach called the Transfer Function Model (TFM) was given by Jury [1982] and is reminiscent of impulse-response formulations in systems analysis. The working hypothesis was that processes in real soil are too complicated for deterministic modeling, and that density functions for the measured travel time may accurately characterize transport. For simple initial and boundary conditions travel time distributions may be measured and used to calculate the response for more complicated situations using a convolution integral.

A disadvantage of this approach is that the distribution needs to be assessed for each soil and depth, and that little understanding is gained of the phenomena involved, in agreement with the working hypothesis. A similar model was developed by Rinaldo and Marani [1987]. Their macroscopic mass response function (MRF) was based on a macroscopic first order mass transfer rate between mobile and immobile phases. For an instantaneous input of solute at zero time the MRF is the product between the travel time PDF and a continuous function when the characteristic time of flow is much smaller than of sorption. This work is highly significant as it shows the way to deal with watershed responses in terms of the instantaneous unit hydrograph and a function that may be fitted for particular solutes. Obviously, on the watershed scale the experimental work needed to evaluate that function is more worthwhile from a manager's point of view than for the column scale (involving a large number of columns). For the watershed scale the MRF-approach may therefore prove to be a very useful management tool.

Most work on spatial variable transport has been concerned with non-reactive transport. Transport of reactive solutes taking randomness of flow, solute input and sorption (retardation) into account was considered by Van der Zee and Van Riemsdijk [1987]. For non-linear sorption, such as the Freundlich equation ($0 < n < 1$)

$$S = kc^n \tag{4}$$

the neglection of pore scale dispersion was shown to be more justified than for tracers that do not react or adsorb linearly. They analyzed the front of saturation (instead of concentration) to allow the use of for non-linear sorption and complicated boundary conditions. Using (3) a simple closed form solution was derived which for linear sorption can be written in terms of the field average concentration front ($\bar{c}$) as

$$\bar{c}(z,t) = \frac{1}{2}\left\{1 - \operatorname{erf}\left[\frac{\ln(z)-m_{\ln Y}}{s_{\ln Y}\sqrt{2}}\right]\right\} \tag{5}$$

[The statistics of the dimensionless penetration depth $Y=vt/RL$ depend on the statistics of v, R, and L and were assumed lognormally distributed. Scaling theory for flow is easily incorporated and used by Van Ommen et al. [1989] to describe bromide transport in a variable field. Although a reasonable description was obtained, the variability in the latter field may have been due to other mechanisms (such as fingering) than are in agreement with the similar media concept. Incorporation of suitably scaled sorption relations and correlation of v and R is also straightforward [Van der Zee and Van Riemsdijk, 1987].

To illustrate some phenomena of transport in spatially variable soil systems, the displacement is considered for a solute subject to linear reversible adsorption and a first order irreversible transformation. Neglecting pore scale dispersion, the transport equation for a particular homogeneous column is

$$R\frac{\partial c}{\partial t} = -v\frac{\partial c}{\partial z} - \mu c \tag{6}$$

For simplicity, a stepwise increase from the zero initial resident concentration to c_0 at time $t = 0$ is assumed. The solution of (6) is

$$c(z,t) = 0; \qquad t \leq Rz/v \tag{7a}$$

$$c(z,t) = c_0 \exp\left(-\frac{\mu z}{v}\right) \qquad t > Rz/v \tag{7b}$$

In a first approach which is justified later only the hydraulic properties are assumed to be random, with the scaling parameter lognormally distributed. The flow velocity and the residence time of solute in a layer of thickness z are distributed, and the residence time (t_r) is given by

$$t_r = \frac{\theta R z}{K(\theta)} \tag{8}$$

To simplify the analysis it is assumed that $\theta = \bar{\theta}$ is a constant. Then all variability is attributed to variability of the hydraulic conductivity at this particular (time averaged) water content. This constrained situation implies that the infiltration rate differs spatially due to other factors than hydraulic properties (e.g. small differences in topographic heigth). However, it should be noted that for more realistic situations the water potential, the hydraulic conductivity, and the volumetric water fraction are coupled. These situations would require more knowledge regarding the hydraulic behaviour of soil, such as the retention curve or the $K(\theta)$-relationship. Such improvements are, however, not the primary scope of this paper. Then, because $K(\theta) = \alpha^2 K_m$ where $\underline{K_m}$ is a spatial average value, we find that the residence time is distributed lognormally ($v = K(\theta)/\theta$) with the statistics

$$m_{\ln(t_r)} = \ln\left(\frac{\bar{\theta} R z}{K_m}\right) - 2m_{\ln(\alpha)}\ ; \qquad s^2_{\ln(t_r)} = \left(2s_{\ln(\alpha)}\right)^2 \tag{9}$$

Due to the differences in flow velocities we obtain a spatial variable field of concentrations. The spatially averaged concentration distribution as a function of depth (z) is given by (3) and equals for depth (z) and time (t) using the chain rule

$$\bar{c}(z,t) = \int_0^{c_0} \tilde{c}\, dF_c(z,t;\tilde{c}) = c_0 - \int_0^{c_0} F_c(z,t;\tilde{c})\, d\tilde{c} \tag{10}$$

In (10) F is the distribution function that gives the probability of c being smaller than $\tilde{c}$. In words, F gives the probability that c is smaller than $\tilde{c}$ because the front has not arrived yet at a particular depth (z) for some of the locations, summed with the probability that the front did reach the depth (z) but the concentration has already decreased below $\tilde{c}$ due to decay. Equation

(10) can be solved analytically form, keeping the effects of incomplete breakthrough and decay separated because at a particular moment both t and βR are constants, where

$$\beta = \frac{1}{\mu} \ln\left(\frac{c_0}{\tilde{c}}\right) \tag{11}$$

For a lognormal residence time distribution this yields

$$F_c(z,t;\tilde{c}) = \frac{1}{2}\left\{\operatorname{erfc}\left[\frac{\ln(t)-m}{s\sqrt{2}}\right] + H(t-\beta R)\left(\operatorname{erfc}\left[\frac{\ln(R\beta)-m}{s\sqrt{2}}\right] - \operatorname{erfc}\left[\frac{\ln(t)-m}{s\sqrt{2}}\right]\right)\right\} \tag{12}$$

where m and s are the statistics (9) of the residence time. It may be noteworthy that when decay is insignificant, (12) simplifies to the first error function as H = 0 always, which is equivalent to the solution given by Van der Zee and Van Riemsdijk [1987] for spatial variable v (and R), i.e., (5). To evaluate the mean concentration in the discharge to the ground water, at depth z=L, it is be convenient to express the residence time in terms of the discharge. Then, in the previous equations t has to be multiplied by $\overline{K}$, and the residence time t_r is replaced by $\tau = t_r\overline{K}$, , where

$$\overline{K} = K_m \exp\left(2m_{\ln(\alpha)} + 2s^2_{\ln(\alpha)}\right) \tag{13}$$

where, as before K_m is a spatial average value. For this situation we have to replace also $\beta \Rightarrow \beta\overline{K}$ and the statistics of the residence time by those of τ

$$m_{\ln(\tau)} = \ln(\theta RL) + 2s^2_{\ln(\alpha)}, \quad s^2_{\ln(\tau)} = \left(2s_{\ln(\alpha)}\right)^2 \tag{14}$$

as

$$\tau = \frac{\theta RL}{\alpha^2} \exp\left(2m_{\ln(\alpha)} + 2s^2_{\ln(\alpha)}\right) \tag{15}$$

It must be noted, though, that the approximation of leaching fluxes based on the areally averaged concentration and discharge is not correct. As Cvetkovic and Destouni [1989] showed, an improved estimation is based on flux concentrations rather than resident concentrations.

When R and or the decay coefficient are also distributed, the solution becomes more complex, because we have to evaluate a probability that $t - \beta R \leq 0$. This probability is not independent from the probabilities of incomplete breakthrough or of reduced concentrations due to decay. For brevity the closed form solution for variable μ and R is not given. In the remainder of this paper, the above situation is analyzed to understand the behaviour of (12) and the effect of parameter values on this behaviour.

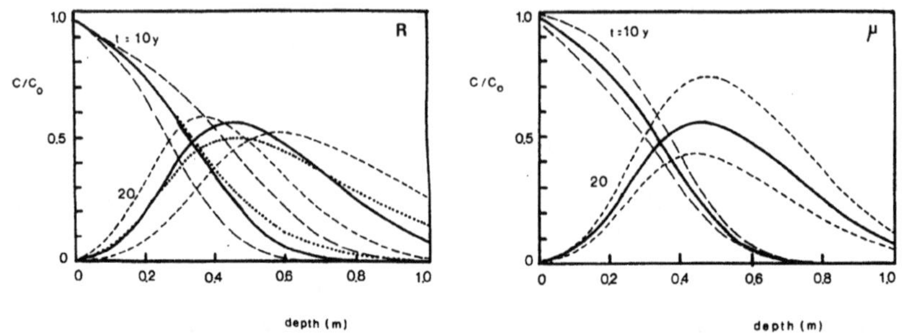

Figure 1: Field average concentration fronts for random R (1a) and random decay (1b):Dotted curves. Also shown the fronts for field equivalent columns with mean values (bold solid lines) and mean plus or minus one standard deviation values (dashed lines) of R (1a) and μ (1b), respectively. Distributions were normal, and the dispersivity was 3 cm.

3. RESULTS: SPATIALLY VARIABLE PESTICIDE TRANSPORT

Pesticide transport is illustrated using parameter values found for atrazine. This substance reacts with the solid phase according to a linear adsorption equation given by $S = f_{om}k_{om}c$.The value of k depends among others on the mass fraction (g/g) of organic matter (denoted f_{om}). For f_{om} = 0.032, a water fraction $\theta = 0.14$ and a dry bulk density of ρ = 1400 kg/m^3, we find with k_{om} = 0.077 a linear retardation factor ($R = 1 + \rho\Delta S/\theta\Delta c$) equal to 25.6. For the same conditions, at a temperature of T = 10 °C, a first order rate coefficient equal to μ = 0.92y^{-1} may be found [Walker, 1978].

In preliminary Monte Carlo calculations, the transport was investigated for mean Dutch conditions (flow velocity v equal to 1.0 m/y). It was assumed that atrazine applications were discontinued after 13 years. The simulations were done by sampling random parameter values from a normal distribution, and solving the analytical solution including dispersion and first order decay, for each sample. The mean concentration at designated depth and time followed from the arithmetic mean. Three parameter values were taken random (separately), i.e., v (CV=0.2), R (CV=0.25), and μ (CV=0.6). Compared with the result for a mean value of these three parameters (the so-called field equivalent column) differences were observed at times of 10, 20 and 30 y, for variability of either v or R. In Figure 1a the effects of variability of R is shown. These two parameters affect the distribution of front depths. Hence, their variability induces a more spread out average solute front. Due to the relatively small coefficients of variation (CV) of v and R, and the relatively large dispersivity, which all agree more or less with measured values, differences are to a large degree averaged away. In Figure 1b the effect of the decay coefficient is illustrated. The field equivalent and the averaged field scale fronts are almost identical. This may be explained with the effect of decay on transport, which does not affect the front penetration depth, but only the concentration level. Due to the field averaging, large concentrations are balanced by relatively small concentrations at the same depth. Profound differences between the field averaged and field equivalent fronts due to variability of decay only, are therefore expected when μ is non-symmetrically distributed (e.g. lognormally, with a large CV).

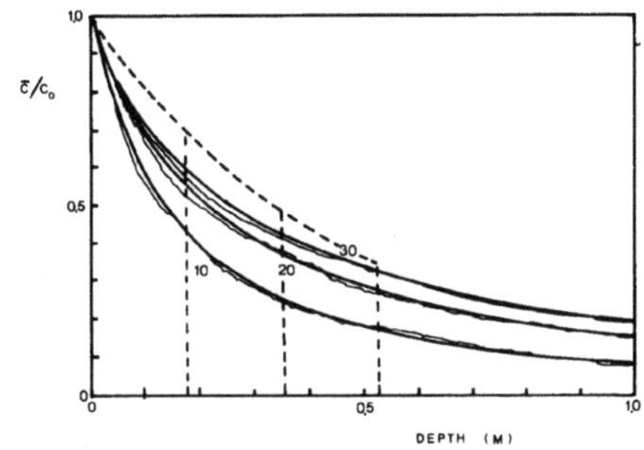

Figure 2: Field average concentration fronts for random flow and times (years) indicated at the curves. Analytical solution (10,12): bold solid line, Numerical solution: thin solid line, Mean field equivalent fronts: dashed lines.

On the basis of the preliminary calculations the decay coefficient was assumed non-random in the remainder of this paper. For continuous atrazine applications (12) may be used. The largest effects compared with the field equivalent column are expected assuming negligible dispersion. The volumetric water fraction was taken 0.14 throughout the field. The saturated hydraulic conductivity, and the statistics for the scaling parameter were given by Van Ommen et al. [1989], and equalled 2.37 m/y (at $\theta_s = 0.316$), $m_{ln(\alpha)} = 0.65$, and $s_{ln(\alpha)} = -0.196$, respectively. In Figure 2 the field averaged front is shown for random hydraulic properties. Monte Carlo simulations and the analytical solution (10, 12) appear to be well in agreement. A gradual increase in the average concentration may be observed as time proceeds from 10 till 30 y. On average, the fronts will have passed the depth of 1 m for times larger than about 26 y, and a steady state front will be approached. Due to variability of flow (and contact times) this steady state front will also be subject to randomness. For illustration, the field equivalent fronts given by (7), using the mean flow velocity v are also shown (Figure 2). It clearly differs significantly from the random fronts, also when dispersion were taken into account. The limiting field average front, with $t \rightarrow \infty$; $t \geq \beta$, maintaining in (12) only the term containing β, is approached slowly by the mean field equivalent front. Variability of water residence times on the *limiting* field averaged front may therefore be neglected for a first estimate.

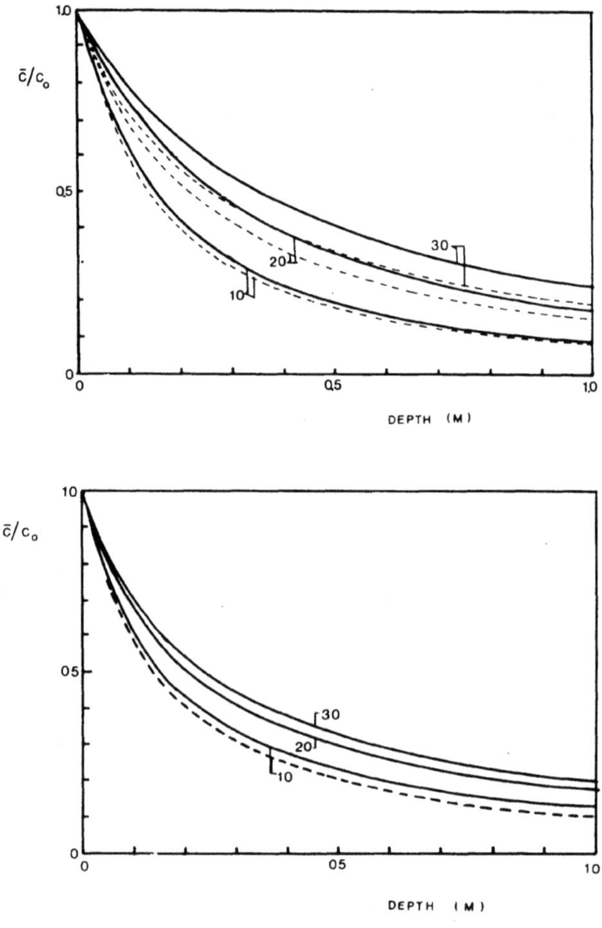

Figure 3: Field average concentration fronts for times indicated in the figure using (10, 12). Figure 3a: effect of decrease of decay coefficient, Figure 3b; effect of increase of R. Reference case: dashed lines, Adapted case: solid lines.

The effects of a different mean decay coefficient and retardation factor are shown in Figure 3a,b. Due to the larger effect of variability of v when decay rates decrease, the lines in Figure 3a are separated more than in the reference simulation, and shifted to larger concentrations. The steady state front is approached equally fast as in the reference case, as the steady state concentration

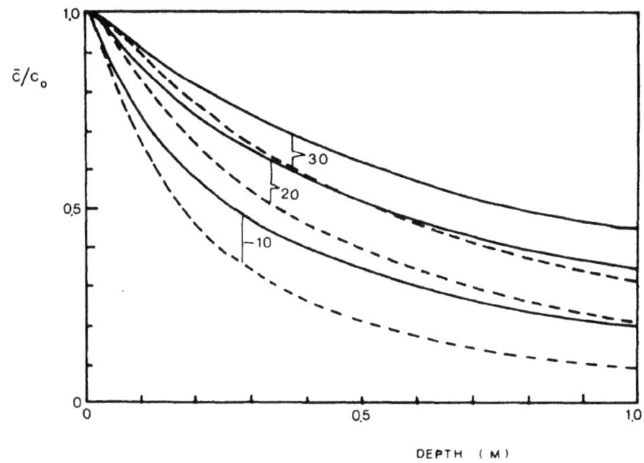

Figure 4: Field average concentration fronts for the case of no decay, with random flow and retardation. CV(R) = 1 : solid lines; CV(R) = 0 : dashed lines.

levels, but not the residence times are affected. This is not the case if R is changed. A two fold decrease in R has a different effect than a twofold decrease in μ and leads to faster breakthrough at designated depths. This causes a faster approach to the limiting steady state front. The steady state front itself is the same as for the reference case where R was 25.6 instead of 12.8, because a single valued R in the terms $R\beta$ (12) and m (see (9)) cancels. For brevity, no effects of variable R, which has a more complicated solution, are presented.

The limiting front is approached faster, i.e., for small times the concentration fronts (CV(R) < 1, mean R=25.6) have slightly higher concentration values than for the reference case with single valued R (Figure 2). The limiting front is the same for single valued or variable R, because R does not show up in the limiting front for each column given by (7). Variability of R is therefore only relevant when most of the columns have not broken through at the lower depth considered. Variability of R had only minor effects even if CV(R)=1. Larger effects are seen when R is random for the case of no decay. In Figure 4 field average fronts with v and R random are shown as obtained with (5). Of interest are the very similar shapes compared with the other curves, which suggests that measurements done in a spatially variable field may give problems in assessing whether first order decay (or uptake) occurs or not. The assumption by Rinaldo and Marani [1987] that on the watershed scale reactivity may be lumped into an overall first order process may be supported by the similarity of the curves of Figures 2-4. Results for random v and R in the absence of decay (not shown) did not affect this last observation.

4. CONCLUDING REMARKS

A model to predict displacement in spatially variable fields was developed, for a solute undergoing adsorption and first order decay. An analytical solution was presented when hydraulic properties are random and lognormally distributed. For the parameter values used (valid for sandy soil and atrazine displacement) variability of flow dominated displacement rather that variability of the decay coefficient or the retardation factor. For very skewed decay coefficient distributions this may not be the case. When the characteristic time for decay becomes much larger than for transport (small μ) variability of R may also have profound effects. Despite the results obtained, further study is awarded to the case of variable v, R and decay parameters. Recognizing the complications involved in pesticide transport I feel that simple solutions are useful to understand the structure of the process of displacement in the field. Thus, unexpected large concentrations of pesticides in some areas in the Netherlands could be explained and understood by taking variability of parameters into account.

The presented analysis was based on flow variability that may be described with the scaling theory of similar media. Due to surface roughness, hydrophobicity of the soil, layering etc. different flow variability may result, that does not agree with this theory. Thus, fingering phenomena may cause highly variable flow and severe bypassing of part of the reactive soil matrix. This type of variability may sometimes be scaled with the similar media theory (with non-realistic parameter values [Van Ommen et al., 1989]), but more research is needed to understand and quantify such flow variability.

NOTATION

c	concentration, c_0 feed concentration (kg/m^3)
$\bar{c}$	field average concentration
CV	coefficient of variation (= s/m)
D	dispersion coefficient (m^2/yr)
F	function
f	frequency distribution
f_{om}	coefficient
h	hydraulic head (m)
K	hydraulic conductivity (m/yr)
k, k_{om}	coefficient (m^3/kg)
L	column length (m)
m	mean
n	constant
R	retardation factor
S	adsorbed amount (kg/kg)

s	standard deviation
t	time (yr)
t_r	residence time (yr)
v	interstitial flow velocity (m/yr)
y	penetration depth (m)
z	depth (m)
α	scaling parameter
β	parameter
τ	volumetric moisture fraction
μ	first order decay coefficient (yr^{-1})
ρ	dry bulk density (kg/m^3)

References

Bresler, E., and G. Dagan, 1981: *Convective and pore scale dispersive solute transport in unsaturated heterogeneous fields*. Water Resour. Res., **17**, 1685-1693.

Cvetkovic, V., and G. Destouni, 1989: *Comparison between resident and flux-averaged concentration models for field scale transport in the unsaturated zone*. In: contaminant Transport in Groundwater, Kobus & Winzelbach (eds.), Balkema, Rotterdam, 245-250.

Dagan, G., and E. Bresler, 1979: *Solute dispersion in unsaturated heterogeneous soil at field scale, I, Theory*. Soil Sci. Soc. Am. J., **43**, 461-467.

Jury, W.A., Simulation of solute transport using a transfer function model, Water Resour. Res., 18, 363-368, 1982

Rinaldo, A., and A. Marani, 1987: *Basin scale model of solute transport*. Water Resour. Res., **23**, 2107-2118.

Van der Zee, S.E.A.T.M., and W.H. van Riemsdijk, 1987: *Transport of reactive solute in spatially variable soil systems*. Water Resour. Res., 23, 2059-2069, 1987

Van Genuchten, M.Th., A closed form equation for predicting the hydraulic conductivity of unsaturated soils, Soil Sci. Soc. Am. J., **44**, 892-898.

Van Ommen, H.C., J.W. Hopmans, and S.E.A.T.M. van der Zee, 1989: *Prediction of solute breakthrough from scaled soil physical properties*. J. Hydrol. **105**, 263-273.

Walker, A., 1978: *Simulation of the persistence of eight soil applied herbicides*. Weed Res., **18**, 305-313.

Warrick, A.W., G.J. Mullen, and D.R. Nielsen, 1977: *Scaling field measured soil hydraulic properties using the similar media concept*. Water Resour. Res., **13**, 355-362.

Sjoerd E.A.T.M. van der Zee, Department of Soil Science and Plant Nutrition, Agricultural University, Dreijenplein 10, 6703 HB Wageningen, The Netherlands

Field-Scale Water and
Solute Flux in Soils
Monte Verità

A PERTURBATION SOLUTION FOR TRANSPORT AND DIFFUSION OF A SINGLE REACTIVE CHEMICAL WITH NONLINEAR RATE LOSS

D.O. Lomen, A. Islas, A.W. Warrick

A perturbation solution is obtained to the partial differential equation describing one-dimensional water and chemical transport in soil systems. Uptake by plant roots is considered in determination of the steady water velocity while the chemical may be lost by plant uptake as well as by first or second-order reactions. Examples illustrate the effect of nonlinear chemical kinetics as well as diffusion.

1. Introduction

The movement and fate of pollutants and agricultural chemicals in unsaturated soil has recently received considerable attention by researchers because of growing environmental concern regarding our soil and ground water. Most of the analytical solutions of partial differential equations used to model these processes require a steady water regime and zero or first order kinetics. However, there are several situations where a nonlinear expression is needed to adequately model the kinetic reaction. Enfield et al. (1982) and van der Zee and Riemsdijk (1987) consider a reaction term of the Freundlich type (Kc^n) where c is the concentration of the solute and K and n constants. Van der Zee and Bolt (1988) (as well as many others) consider Langmuir adsorption where the concentration reactive term has the form $\alpha c/(1 + \beta c)$.

For situations where c is scaled to be less than 1, a Taylor series expansion of the last expression gives $\alpha c(1-\beta c + \ldots) = \alpha c - \alpha\beta c^2 + \ldots$. This paper is concerned with models where the kinetic term has both a linear and quadratic term. It builds on the analytical solution developed by Lomen et al. (1989) for a dispersion-free situation.

Reviews of analytical solutions of linear versions of equations (1) are included in papers by Nielsen et al. (1986) and van Genuchten and Jury (1987). Yates and Enfield (1989) also include a squared term as they solve a pair of partial differential equations modeling a second order reaction.

2. The Governing Equations

The partial differential equation for one-dimensional movement of a reactive chemical in a porous medium is obtained by combining the continuity equation with apparent diffusion as

$$\frac{\partial}{\partial t}(\theta c + \rho_b s) = \frac{\partial}{\partial x}\left[\theta D\frac{\partial c}{\partial x} - qc\right] - \beta_s\rho_b s - \beta_1\theta_c - \mu\theta c^2 - [1-\gamma]cS \ , \tag{1}$$

where θ is the volumetric water content (dimensionless); c is the solute concentration (mol/cm^3); ρ_b is the bulk density of the porous medium (g/cm^3); s is the adsorbed concentration (mol/g); D is the apparent diffusion coefficient (cm^2/d); q is the Darcian flux (cm/d); β_1, μ and β_s are rate constants associated with liquid and solid phases of the chemical (d^{-1}); γ is a "reflection" coefficient associated with solute being taken up by the roots (dimensionless with $\gamma = 0$ is for no reflection and $\gamma = 1$ for total reflection, cf. Hillel (1980), p. 246); S is the plant water uptake function (d^{-1}); t is time (d); x is distance below the soil surface (cm). (See also van Genuchten, 1981.) Reviews of analytical solutions of linear versions of equation (1) are included in papers by Nielsen et al. (1986) and van Genuchten and Jury (1987). Assume we have an equilibrium adsorption state (which may be nonlinear) in the form

$$s = K_d F(c) \tag{2}$$

and that the water movement and uptake by plant roots are governed by one-dimensional flow, so

$$\frac{\partial \theta}{\partial t} = -\frac{\partial q}{\partial x} - S \ . \tag{3}$$

Combining equations (1), (2) and (3) gives

$$(\theta + \rho_b K_d F'(c))\frac{\partial c}{\partial t} = \frac{\partial}{\partial x}(\theta D\frac{\partial c}{\partial x}) - q\frac{\partial c}{\partial x} - \left[\beta_1\theta + \beta_s\rho_b K_d\frac{F(c)}{c} - \gamma S\right]c - \mu\theta c^2 \ , \tag{4}$$

where $F'(c) = dF/dc$. In order to solve equation (4) for the chemical concentration (c), the function describing the equilibrium adsorption state, equation (2), must be given as well as the values of θ, q, and S which describe the waterflow. The solution of equation (4) is now presented for three special cases. The values of S, q, and θ are for steady flow and listed in Table I (Warrick, 1974) and (Lomen and Warrick, 1976). These papers use a hydraulic conductivity function of the form $K = K_0 \exp(\alpha h)$ and $\theta = (K - K_0)/k + \theta_{sat}$.

Table I: Values of S, q, θ for steady flow.

Exponential sink
$S = ae^{-bx}$
$q = q_0 + (e^{-bx} - 1)a/b$
$\theta - \theta_{sat} = (q_0 - a/b)/k - K_0/k + \alpha a e^{-bx}/[kb(b + a)]$
Step sink
$S = \begin{cases} a_1 , & 0 < x < L_1 \\ a_2 , & L_1 < x < L_2 \\ 0 , & L_2 < x \end{cases}$
$q = \begin{cases} q_0 - a_1 x , & 0 < x < L_1 \\ q_0 - a_1 L_1 - a_2(x - L_1) , & L_1 < x < L_2 \\ q_0 - a_1 L_1 - a_2(L_2 - L_1) , & L_2 < x \end{cases}$
$\theta - \theta_{sat} = \begin{cases} (\alpha(q - K_0) - (a_2 - a_1)e^{\alpha(x - L_1)} + a_2 e^{\alpha(x - L_2)} - a_1)/\alpha k , & 0 < x < L_1 \\ (\alpha(q - K_0) + a_2 e^{\alpha(x - L_2)} - a_2)/\alpha k , & L_1 < x < L_2 \\ (q - K_0)/k , & L_2 < x \end{cases}$

3. Example I: No diffusion with a linear reaction or adsorption

The perturbation solution of equation (4) for the nonlinear case and for the case including diffusion rely on that for the linear case with no diffusion. Thus we let $D = \mu = 0$ and $F(c) = c$ in equation (4) and consider a surface boundary condition representing a pulse application at the surface,

$$c(0,t) = A_0 \, [H(t) - H(t - t_0)] \ , \tag{5}$$

where $H(t)$ is the Heaviside function (0 for $t < 0$, 1 for $t > 0$). The solution for zero initial concentration is

$$c(x,t) = A_0 e^{-f_2(x)} \, [H(t - f_1(x)) - H(t - t_0 - f_1(x))] \ , \tag{6}$$

where (Lomen et al., 1984)

$$f_1(x) = \int_0^x \frac{\theta(y) + \rho_b K_d}{q(y)} \, dy \ , \tag{7}$$

$$f_2(x) = \int_0^x \frac{\beta_1 \theta(y) + \beta_s \rho_b K_d - \gamma S(y)}{q(y)} \, dy \ , \tag{8}$$

From the form of equation (6) we see that the chemical concentration for a "pulse" application at the soil surface is a "pulse" moving downward into the soil. The chemical "pulse" will first arrive at depth x at time $f_1(x)$ and completely pass by at time $f_1(x) + t_0$. The "shape" of the profile is given by $A_0 \exp[-f_2(x)]$.

4. Example II: No diffusion with a nonlinear reaction or adsorption

The pertinent partial differential equation for this case ($D = 0$) is

$$\frac{\partial c}{\partial t} = -V(x)\frac{\partial c}{\partial x} - K(x)c - \mu c^2 , \tag{9}$$

where

$$K(x) = \frac{\beta_1 \theta(x) + \beta_s \rho_b K_d - \gamma S(x)}{\theta(x) + \rho_b K_d} ,$$
$$V(x) = \frac{q(x)}{\theta(x) + \rho_b K_d} . \tag{10}$$

For arbitrary surface concentration, $c(0, t) = \psi(t)$, and zero initial concentration, $c(x, 0) = 0$, we may derive (using the method of characteristics) a solution of the form

$$c(x,t) = \frac{\psi(t - f_1(x))e^{-f_2(x)}}{1 + \mu\psi(t - f_1(x))f_3(x)} \tag{11}$$

with

$$f_3(x) = \int_0^x V^{-1}(y)e^{-f_2(y)}dy . \tag{12}$$

We give two examples now to illustrate the effect of the nonlinear term. Figure 1A pertains to the exponential sink of Table I. Here we see the solution for $\mu = 0, 0.01, 0.015$, and 0.02. The solution for $\mu = 0$ is the one with the largest concentration and is from Lomen et al., 1984. Figure 1B uses the step sink of Table I and the same values of μ. Again the $\mu = 0$ curve corresponds to the largest concentration.

5. Example III: Small diffusion with a linear reaction

In this case $D << 1$ and $\mu = 0$, so the appropiate equation is given by

$$\frac{\partial c}{\partial t} = D\frac{\partial^2 c}{\partial x^2} - V(x)\frac{\partial c}{\partial x} - K(x)c . \tag{13}$$

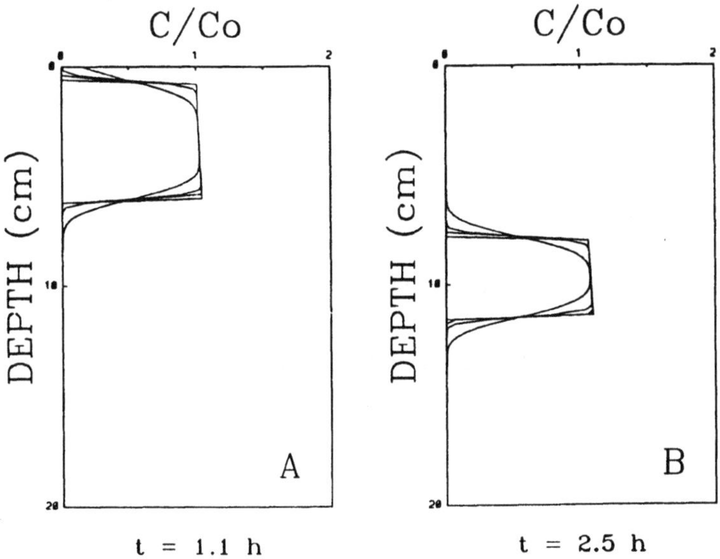

Figure 1. Comparison of concentration profiles for four values of D (0, 0.1, 0.01, 0.001) for a step input and a step sink at times 1.1 h (A) and 2.5 h (B).

Since the small parameter D is the coefficient of the highest partial derivative we have to use the method of singular perturbations. For the *outer solution,* we set $D = 0$, so the solution is given by equation (6) ($t_0 = \infty$). For the *inner solution,* we assume $D \ll 1$, and make the change of variables

$$\xi = \frac{x - y(t)}{D} \quad , \quad U(\xi,t) = c(y(t) + D\xi,t) \ ,$$

where $y(t)$ is the location of the "jump" (yet to be determined).
This gives

$$\frac{\partial^2 U}{\partial \xi^2} - (V(y(t) + D\xi) - y'(t))\frac{\partial U}{\partial \xi} = D\left[\frac{\partial U}{\partial t} + K(x)U\right] , \tag{14}$$

where $y'(t) = dy/dt$. We solve this equation for the "jump" at the leading edge of the pulse ($y = y_1$), as well as the trailing edge ($y = y_2$).

Omitting many details [Islas et al., 1990] we obtain our final expression for chemical concentration as

$$c(x,t) = c_0(x,t) + U_1(\xi,t) - c_0(y_1,t)H(x-y_1) + U_2(\xi,t)c_0(y_2,t)H(y_2-x) \, , \tag{15}$$

where

$$U_1(\xi,t) = c_0(y_1,t)\int_{-\infty}^{\xi} e^{DV'(y_1)\eta^2/2}d\eta / I(y_1) \, ,$$

$$U_2(\xi,t) = c_0(y_2,t)\left[1 - \int_{-\infty}^{\xi} e^{DV'(y_2)\eta^2/2}d\eta / I(y_2)\right] ,$$

$$I(y) = \int_{-\infty}^{\infty} e^{DV'(y)\xi^2/2}d\xi \, ,$$

where $V'(y) = dV/dy$. Figures 2A and 2B show the solution from equation 15 for the step sink of Table I at two different times. The "square curve" is for $D = 0$ and the deviations from this square increases as D increases.

6. Conclusion

While the main purpose of this paper is to present a perturbation solution to the partial differential equation modelling dispersion of a solute in a porous medium with nonlinear kinetics, it is far from just a mathematical exercise. While most models of solute movement at the field scale use zero or first order terms, as understanding of these processes increases and mathematical solutions are developed, more complicated mathematical models will be considered. We note that van der Zee and Riemsdijk (1987) used a Freundlich model to compare with their field experiments. In addition to direct applications, such solutions also serve as valuable checks for numerical algorithms developed for even more general situations.

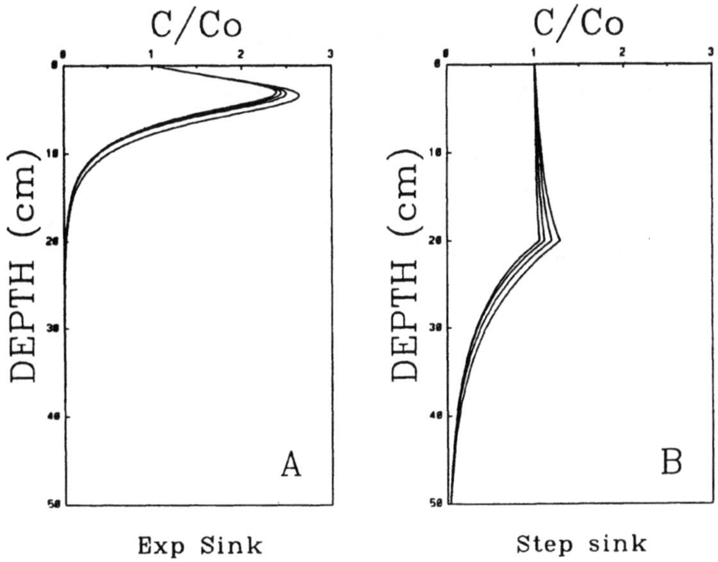

Figure 2. Comparison of concentration profiles for four values of μ (0, 0.01, 0.015, 0.02) for an exponential sink (A), and a step sink (B).

References

Enfield, C.G., R.F. Carsel, S.Z. Cohen, T. Phan, and D.M. Walters,1982: *Approximating pollutant transport to ground water*. Ground Water, **20**, 711-722.

Hillel, D. 1980: *Fundamentals of Soil Physics*. Academic Press, New York, NY.

Islas, A., D.O. Lomen, and A.W. Warrick, *A perturbation solution of the transport equation for small diffusion and nonlinear rate loss*. (in preparation)

Lomen, D.O., P.J. Tonellato, and A.W. Warrick, 1984: *Salt and water transport in unsaturated soil for non-conservative systems*. Agricultural Water Manage, **8**, 397-409.

Lomen, D.O., A. Islas, X. Fan, and A.W. Warrick, 1990: *A perturbation solution of a nonlinear transport equation for a porous media*. Trans. in Porous Media (accepted).

Lomen, D.O. and A.W. Warrick, 1976: *Solution of the one-dimensional linear moisture flow equation with implicit water extraction functions*. Soil Sci. Soc. Am. J., **40**, 342-344.

Nielsen, D.R., M.Th. van Genuchten, and J.W. Biggar, 1986: *Water flow and solute transport processes in the unsaturated zone*. Water Resour. Res., **22**, 895-1085.

van der Zee, S.E.A.T.M., and W.H. van Riemsdijk, 1987: *Transport of reactive solute in spatially variable soil systems*. Water Resour. Res., **23**, 2059-2069.

van Genuchten, M.Th., 1981: *Analytical solutions for chemical transport with simultaneous adsorption, zero-order decay*. J. Hydrol., **49**, 213-233.

van Genuchten, M.Th. and W.A. Jury, 1987: *Progress in unsaturated flow and transport modeling*. Rev. of Geophys., **25**, 135-140.

Warrick, A.W., 1974: *Solution to the one-dimensional linear moisture flow equation with water extraction*. Soil Sci. Soc. Am. Proc., **38**, 573-576.

Yates, S.R. and C.G. Enfield, 1989: *Transport of dissolved substances with second-order reaction*. Water Resour. Res., **25**, 1757-1762.

D. O. Lomen, A. Islas, Department of Mathematics, University of Arizona, Tucson, AZ 85721.

A.W. Warrick, Soil and Water Science, 429 Shantz #38, University of Arizona, Tucson, AZ 85721

Field-Scale Water and
Solute Flux in Soils
Monte Verità

AREAL SOLUTE FLUX ESTIMATION: LEGAL ASPECTS

G. Karlaganis and J. Dettwiler

1. Introduction

The legislators and, subsequently, the state agencies are facing the task of regulating hazardous substances in the environment. The restrictive regulation relates not only to concentration standards of chemicals or organisms defined for water, air, soils but also with respect to the total permissible load entering or leaving the various environmental compartments. The estimation of areal mass fluxes through soils is one of the difficult tasks of environmental regulation. On one hand there is a discrepancy between accessible and obtainable data and on the other hand it is the ultimate goal of minimizing the level of accumulation within a given compartment, i.e. in soils, as well as controlling the fluxes of xenobiotics from one compartment to another, e.g. from soil to groundwater or from atmosphere to soil etc.
The accessibility of relevant soil pollution information is not only limited due to economic reasons but also with respect to our understanding of what should be looked at or what could be done with soil data.
It is common practice, not only in Switzerland or Central Europe, to rely on pollutant concentrations as measured by certain more or less standardized extraction procedures. It is a crucial point that such information should not only enable us to assess the actual pollution level but also to predict the time frame of the contamination and decontamination processes. In this context, we would like to contrast the legal base and current practices of preventing soil and groundwater contamination in Switzerland with the state of the art of solute transport prediction in field soils.

2. The Legal Base

Switzerland has a population of approximately 6.8 million inhabitants and covers a territory of some 41'000 km^2. Roughly a quarter of this surface, namely the plain lowlands ("plateau") together with the large bottom valleys within the Alps, is densely populated. The Jura ranges and the mountainous areas of the Northern and Southern foothills of the Alps account for another quarter of the territory. Forests and unproductive land—lakes and mountains—make up the other half of the country. Thus, the major portion of our industrial and agricultural activities which have an impact on the environment are confined to only one quarter of the country's surface. The signs of increasing environmental pollution which can be attributed primarily to these activities as well as to traffic became unmistakably obvious. The density of population in this area is about 500 inhabitants per km^2.

Swiss environmental legislation is based on two main laws aimed (among others) at limiting the use and emissions of environmentally hazardous substances. The two laws are the Federal water pollution control act of 1971 and the Federal law for environmental protection of 1983. As a result, pollution preventing measures, some very strict, are presently being implemented on the level of the pollution sources (emission control) as well as on the level of pollutant recipients (immission control), i.e. water, air, soil and indirectly the population, animals and plants. Moreover, the law on foodstuffs and the law on toxic substances allow for legal measures to be taken for protecting the consumer.

The legal measures for restricting emissions of hazardous substances into air, water and soil are implemented in accordance with the corresponding ordinances of the Federal Council i.e. the clean air act (1985), the ordinances on sewage sludge (1981) and on waste water discharge (1975).

In the ordinance on soil pollutants (1986), the Federal Council moreover defined the country-wide principals of evaluating soil pollution levels and initiated a program for monitoring the soil pollution for 10 heavy metals and for fluoride, the so-called soil immision reference values (environmental standards) (Table 1). Even after prolonged accumulation periods the pollutant concentrations in soils should neither reach nor exceed these standards given in Table 1. Common concentrations of heavy metal contents found in Swiss soils originate partly from anthropogenic as well as from soil mineral sources. In some areas the concentrations of certain heavy metals significantly exceed the reference values. A well known example is the high copper contents in vineyards.

3. The National Soil Pollution Surveillance Network (NABO)

The NABO-Network established in 1984 is intended to observe the longterm trends of non-point source pollution. At one hundred selected sites the contents of the eleven key elements mentioned in Tab. 1 are periodically monitored according to an established analytical guide-line. Additional elements may be added to the monitoring program in the future. The sampling sites were carefully selected to represent a wide spectrum of land uses and soil varieties in agricultural areas, agglomerations as well as in "pristine" areas. The first sampling of the NABO-sites provides a base-line data set to be used as a frame of reference for more detailed studies by cantonal authorities in problem areas where a contamination with specific pollutants can be assumed or has already been proven.

Table 1: Concentrations of heavy metals in Swiss soils

Elements	Common Concen-trations ppm	Reference values ppm	Known Pollutant Emitters (Concentration ppm)
Pb	0.1-20	50	Highways (200)
Cr	2 - 50	75	Sewage sludge e.g. tanneries
Cd	~ 0.1	0.8	Composts/Industrial areas (1-6)
Co	1 - 10	25	Industrial areas
Cu	1 - 20	50	Vineyards (1500) Industrial areas (2400)
Mo	1 - 5	5	Fertilizers from wastes
Ni	12 - 40	50	Industrial areas (250)
Hg	0.1	0.8	
Ti	0.1-0.5	1	Concrete works
Zn	20-100	200	Industrial areas (>1500)
F	100-400	400	Industrial areas (> 600)

4. Enforcement Responsibilities

When the standards listed in Table 1 are exceeded for one or more element, or, when the rate of accumulation noticeably increases, or, when the fertility of a particular soil can no longer be maintained on its normal level, the Cantons (States) must investigate the cause of decline of soil fertility and the source of pollution.

These legal actions are not restricted to agricultural land and soils directly affected by a point source pollution, but to the entire soil surface of the Swiss territory. The Cantons are obliged to verify the effectiveness of the measures taken in accordance with the federal emission control ordinances—as mentioned above—and to supplement the investigations as well as the preventive or corrective actions according to the particular needs.

5. Soil Fertility and the Legal Base for Soil Protection

In cases of pollutants not listed explicite in the federal ordinance, the impairment in soil fertility is assessed based on its definition given in the ordinance. Thus, a soil is fertile:

a) if its fauna, flora and plant community is biologically active, diverse, exhibiting a typical structure and sustainable decomposition potential;
b) if it maintains normal plant growth and development of both natural and anthropogenic plant communities with unchanged characteristic properties;
c) if it warrants that plant products are of a good quality and wholesome for man and animals.

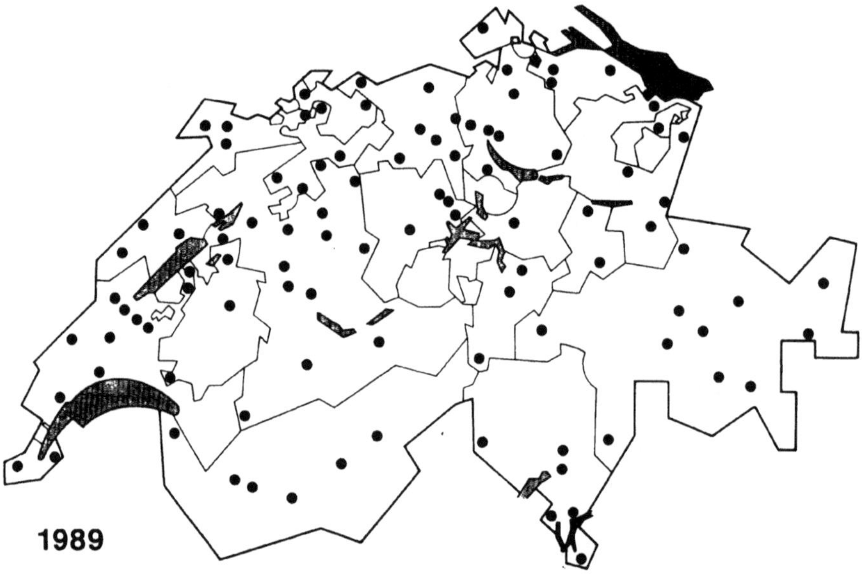

Figure 1: Swiss Soil Surveillance Network "NABO"

6. Mass Fluxes and Law Enforcement

At this point we would like to illustrate the link between field scale solute fluxes and legal measures taken in the context of the water protection law. We focus on phosphorus fluxes which is an important issue in Switzerland (Figure 2). The fluxes are expressed in kilotons per year. The flux diagram on its left shows the cattle and plant production section and on the right the consumer section. The plant production yields a net output of 20 $kt \cdot y^{-1}$ of phosphorus. Approximately 1–2 % of phosphorus contained in the liquid manure (slurry) is estimated to exit into surface waters. These 0.3 to 0.6 $kt \cdot y^{-1}$ of phosphorus induce a considerable eutrophication, which accelerates oxygen consumption in lake waters and toxin production by excessive algae growth and, as a consequence, may cause frequent fishkills.

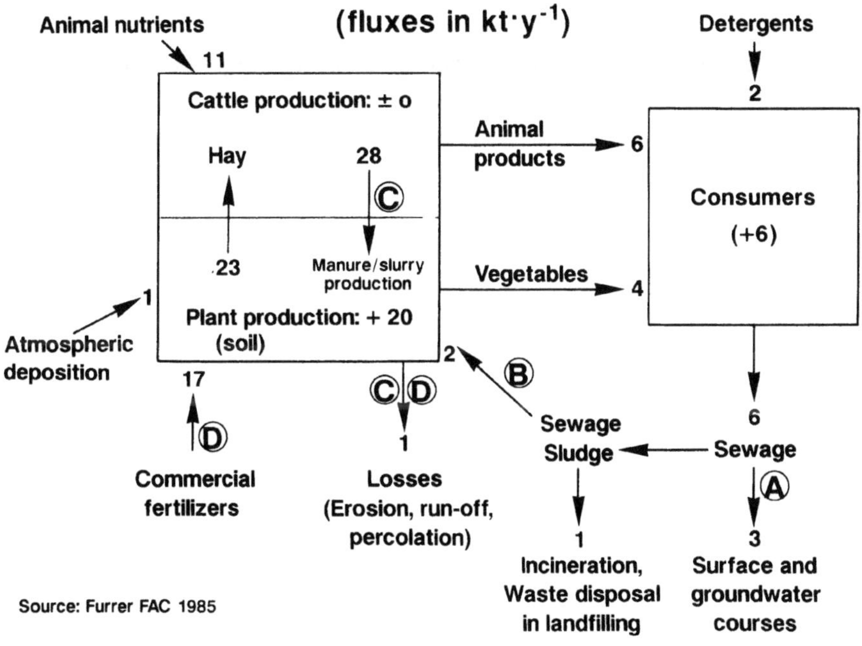

Figure 2: Phosphorus Fluxes in Switzerland

The P-fluxes labelled with letters A to D (Figure 2) refer to regulated areas. Letter A corresponds to the ordinance on waste water discharge of 1975, B to the ordinance on sewage sludge of 1981, C to the measures on slurry production and D to the use of commercial fertilizers. Letters C and D are regulated in the ordinance on environmentally hazardous substances of 1986. Thus, legal measures were taken to regulate solute fluxes onto and through soils.

At the present time these legal measures are enforced based on the amendment of the water protection law. An example of such legal action is the limitation of the areal animal density. Manure of not more than 3 dairy cows—or an equivalent number of other animals which produce the same quantity of manure—will be allowed per hectare. Thus, the peak loads of slurry production—letter C—are cut back, both on an areal as well as temporal base.

Many questions regarding the prevention of soil contamination and conservation of soil fertility are still pending, both from the scientific as well as from the legislative point of view. Long term behaviour of heavy metals, corroborated and operational biological indicators for soil quality estimation, legislation on soil erosion and soil compaction control, or feasible soil decontamination and reclamation procedures are but a few of the problems in this field.

G. Karlaganis and J. Dettwiler, Federal Office of Environment, Forests and Landscape, CH-3003 Berne, Switzerland

MIX
Papier aus verantwortungsvollen Quellen
Paper from responsible sources
FSC® C105338

If you have any concerns about our products,
you can contact us on
ProductSafety@springernature.com

In case Publisher is established outside the EU,
the EU authorized representative is:
Springer Nature Customer Service Center GmbH
Europaplatz 3, 69115 Heidelberg, Germany

Printed by Libri Plureos GmbH
in Hamburg, Germany